Visual Attention-Related Processing

Visual Attention-Related Processing: Perspectives from Ageing, Cognitive Decline and Dementia

Editors

Andrea Tales
Claire J. Hanley

MDPI • Basel • Beijing • Wuhan • Barcelona • Belgrade • Manchester • Tokyo • Cluj • Tianjin

Editors
Andrea Tales
Swansea University
UK

Claire J. Hanley
Swansea University
UK

Editorial Office
MDPI
St. Alban-Anlage 66
4052 Basel, Switzerland

This is a reprint of articles from the Special Issue published online in the open access journal *Brain Sciences* (ISSN 2076-3425) (available at: https://www.mdpi.com/journal/brainsci/special_issues/Visual_Attention).

For citation purposes, cite each article independently as indicated on the article page online and as indicated below:

LastName, A.A.; LastName, B.B.; LastName, C.C. Article Title. *Journal Name* **Year**, *Volume Number*, Page Range.

ISBN 978-3-0365-0984-6 (Hbk)
ISBN 978-3-0365-0985-3 (PDF)

Contents

About the Editors

Claire J. Hanley (B.Sc., MRes, Ph.D.) is a Senior Lecturer in Cognitive Neuroscience & Ageing at Swansea University. Having conducted cutting-edge research at the Centre for Ageing and Neuroscience at Cambridge University (Cam-CAN), before completing her Ph.D. in Advanced Neuroimaging Methods at Cardiff University's Brain Imaging Research Centre (CUBRIC), she has considerable experience of designing studies in neural plasticity, incorporating neuroimaging techniques and non-invasive brain stimulation methods to determine how the brain adapts as part of the ageing process.

Andrea Tales (B.Sc. Hons, M.Sc. (Oxon), Ph.D., FBPsS, DCR[T]) holds a Personal Chair in Neuropsychology and Dementia Research and is a Fellow of the British Psychological Society. Andrea has published extensively in peer reviewed-journals and runs an extensive ageing and dementia-related research network. After completing her Ph.D. in Early Visual processing (visual mismatch negativity) in Alzheimer's Disease, she has had considerable research experience using multi-methodologies to investigate the integrity of visual attention-related brain function in subjective and mild cognitive impairment and various aetiologies of dementia.

Editorial

Visual Attention-Related Processing: Perspectives from Ageing, Cognitive Decline and Dementia

Claire J. Hanley [1,*] and Andrea Tales [2]

[1] Department of Psychology, Swansea University, Swansea SA2 8PP, UK
[2] Centre for Innovative Ageing, Swansea University, Swansea SA2 8PP, UK; A.Tales@swansea.ac.uk
* Correspondence: C.J.Hanley@swansea.ac.uk

Regarded as a defining factor in resource management, it is widely accepted that visual attention and related processing will deteriorate, in a global fashion, across the lifespan and produce detrimental consequences for environmental interactions. Disproportionate detriments in such functions are also proposed to arise in the context of a range of age-related disorders. However, the nature and extent of such losses, the precise mechanisms that underlie these changes, and the origin of deficits in task performance that are used to assess this multicomponent process are still under debate. This Special Issue, comprising studies spanning healthy ageing, cognitive impairment, and dementia, represents a body of literature that fundamentally advances these points of discussion.

Contrary to previous findings and prior assumptions, the research documented herein attests to a large degree of preserved task performance in ageing. First, with regard to the role of attention in working memory (WM), by simultaneously assessing participants' ability to filter and ignore distracting stimuli, Maniglia and Souza [1] separate the roles of encoding and maintenance in the first study to do so in an older adult sample. Through Bayesian analysis, the authors establish a lack of age-related deficit with regard to both components while also providing evidence for similar effects of memory load and distraction across the adult lifespan. These findings signify that older participants had no significant problems allocating sufficient attentional resources to manage visual WM representations compared to their younger counterparts. Although they require extended time to engage attention, older adults were just as efficient at encoding relevant stimulus features. In this instance, the findings suggest that visual WM performance cannot be explained by deficits in attentional control, as is often speculated. Therefore, age-related differences should not be assumed in this context, and where they are observed, should not be automatically presumed to arise as a result of deficient inhibition of distracting items or irrelevant stimulus attributes.

Henderson et al. [2] also identified similar strategies between younger and older adults, suggesting the ability to flexibly manage attention allocation to support visual WM performance is generally maintained across the lifespan. Each group successfully utilised spatial cues to modify attention and ignore distractors when cue validity was completely certain. However, the presence of this implied age-related functional integrity varied depending on the priority assigned to the stimulus. Accordingly, the authors suggest the successful updating of details that are no longer considered relevant is highly influenced by the extent to which older adults have to divide their attention under high demands. The most compelling models of the findings indicated that the high proportion of nontarget errors demonstrated by older adults could be attributed to the lasting impression of the priority assigned at the encoding stage. Therefore, older adults failed to inhibit the no-longer-relevant yet originally prioritised item because initially low-priority items could not be retrieved when resources were in short supply. This proposal has valuable applications in assessing the efficacy of visual WM in older adults, which could be predicted on the amount of resources allocated to a stimulus and may translate to other executive tasks.

Citation: Hanley, C.J.; Tales, A. Visual Attention-Related Processing: Perspectives from Ageing, Cognitive Decline and Dementia. *Brain Sci.* **2021**, *11*, 206. https://doi.org/10.3390/brainsci11020206

Received: 4 February 2021
Accepted: 5 February 2021
Published: 9 February 2021

On the basis of this evidence, where age-related deficits in visual WM are observed (most likely due to priority biases), they should not be attributed to poor strategy formation but, instead, the inability to engage adequate inhibitory control.

Further extending the discussion of integrity of function to those with mild cognitive impairment, Revie et al. [3] highlight an absence of age-related information processing deficits via a drift rate model applied to choice reaction time data. Healthy older adults were found to be more cautious with regard to decision making but did not demonstrate differences in drift rate to those with mild cognitive impairment in Lewy body disease (MCI-LB). Furthermore, this study addresses the debate as to whether the origin of visual hallucinations in those with MCI-LB can be attributed to perceptual or attentional deficits, an important consideration, as hallucinations impact upon cognitive decline and quality of life. The authors conclude that while not significantly impaired, all patients exhibited a slowing of sensory processing, whereas those who experience hallucinations also displayed executive deficits in decision-making. Such knowledge will undoubtedly pave the way for further exploration into the onset of hallucinations as well as targeted interventions to alleviate their detrimental consequences.

Across the lifespan, and between states of health and disease, the extent to which individuals are able to adequately manage attentional allocation remains contested. Extending the debate on the control of resources to the domain of divided attention, Barel and Tzischinsky [4] investigate object–location memory, an ability crucial to everyday functioning. A study with an incredibly novel focus, the authors provide a targeted demonstration of how top-down attention can determine memory performance. Addressing the role of attentional and executive functioning in incidental compared to intentional episodic memory encoding, they demonstrate superior performance in the incidental condition. Furthermore, vigilance correlated with memory performance scores in the incidental condition only, suggesting that under increased cognitive load sustained attention influenced incidental but not intentional encoding.

Additionally, Polden et al. [5] compared the ability of healthy younger and older adult controls to those with MCI and Alzheimer's disease (AD) to successfully disengage and refocus attention, while incorporating eye-tracking to facilitate measures of the "gap effect". With regard to the influence of age, in contrast to their younger counterparts, older participants demonstrated slower reaction times and greater difficulty in disengaging attention, in addition to a general slowing in prosaccades. Perhaps surprisingly, performance and prosaccades were similar in comparison to the patient groups, providing insight into preservation of the "gap effect" in states of cognitive impairment. The authors established that only in the MCI group were longer reaction times correlated with poorer working memory, inferring that such scores—reflecting attentional disengagement—could act as a valuable metric during what is considered to be a transition period to AD for some individuals. The research emphasises that while the superior colliculus appears to remain intact, executive regions of the network such as the anterior cingulate, which are crucial for inhibitory control, are likely to be affected early on in the disease process. Indeed, there may be important lessons to be learned from this study to focus on identifying the subtleties of preserved function instead of fixating on targeting deficits, while also addressing individual differences in factors such as ethnicity. These considerations could have important implications with regard to the assessment of attention-related processing and associated approaches to enhance existing function.

Building on the role of participant characteristics, Torrens-Burton et al. [6] explored the occurrence and origin of age-related attentional control deficits to a complex yet naturalistic task-switching paradigm. Investigating a range of variables (including subjective memory function, cognitive reserve, gender, and education), the authors echo the sentiment of this collection of articles by demonstrating that while age can be inferred to influence speed, it seldom fundamentally alters proficiency. The ability to modulate attention to attend to a target and inhibit interference, under conditions of increasing difficulty brought about by greater task speed, can be considered significantly less efficient in older adulthood

However, age itself was not a contributing factor to performance accuracy, and while statistically slower than younger adults, older participants were no less precise. Scores reflecting general cognition (determined by the Montreal Cognitive Assessment, MoCA) but not age itself, were associated with speed of performance. MoCA performance can, therefore, be considered a potential marker of information processing capacity in the context of the task, but the overarching message is that attentional control does not appear to vary by age per se. Consequently, translating performance of a dynamic task to real-life demonstrations of such executive function (e.g., in the context of driving), the study implies that chronological age is unlikely to be a direct contributing factor in proficiency. Moving forward, it will be of extreme importance to identify the factors that underlie superior cognitive scores (as the MoCA results did not relate to perceived level of cognitive function or the other variables assessed as part of the research) if they are to prove informative.

The real-world insights advocated by the aforementioned study could produce incredibly valuable applications considering driving cessation is a major life transition, with implications for health and well-being. Huizeling et al. [7] enhance our understanding of the complexities of refocusing attention in such a naturalistic setting, presenting the first study to acquire task-related oscillatory data, using electroencephalography (EEG), from a subgroup of older adult participants to extend inferences based on simulated driving measures. The data support the pattern of age-related slowing of reaction times shown consistently in this collection of articles—in this instance, reflecting adjustments older adults make to their driving behaviour to accommodate deficient attentional control mechanisms. Consequently, the processing difficulties reported in abstract computer tests also appear to translate to such realistic settings. The incidence of slow reactions with increased age was established in several tasks: from the time to read road sings to that taken to indicate. Additionally, where attention was already engaged to monitor the behaviour of other drivers, it was more difficult for older adults to widen their focus to identify a relevant target among distractors. These behavioural findings were accompanied by age-related differences in theta and alpha power, which culminate to suggest that older, compared to younger, adults appear to be less able to engage the attentional resources required to switch focus and respond to dynamic changes in a flexible manner. The authors note that switching was affected by how predictable a stimulus was (corresponding to the findings outlined here [1,2]), yet the demands of driving are anything but predictable. Where older adults appear to adopt fundamentally different, and largely maladaptive, strategies during driving, this study provides pivotal evidence of the need for interventions that can facilitate safe driving behaviour, which will be particularly poignant past the point where the bounds of cognitive and neural compensation mechanisms are exceeded (in those most vulnerable to decline; e.g., states of cognitive impairment, lower cognitive reserve, as well as "old-old" age and beyond).

In recent years, non-invasive brain stimulation techniques (NIBS) have been heralded as compensation aids due to their ability to enhance neural plasticity. Use of these methods could, therefore, represent a powerful approach to sustaining cognition in ageing. However, beyond initial successes, the multitude of stimulation parameters available to researchers, especially those exploring transcranial direct current stimulation (tDCS), makes it extremely difficult to optimise protocols and maximise beneficial outcomes. Hanley et al. [8] address this quandary by systematically investigating stimulation duration, using a variation of the dynamic task-switching paradigm that features here [6]. Sham control stimulation failed to enhance performance, suggesting a lack of practice or placebo effects, whereas superior performance speed was documented following active tDCS. However, contrary to predictions, ten minutes of stimulation was shown to produce advantageous results when compared to the longer twenty-minute variant. The authors speculate that the success of the shorter intervention can likely be attributed to participant characteristics, which determine the status of grey and white matter integrity and resulting presence of cerebrospinal fluid. While the results of this study represent a hugely exciting prospect for on-going efforts to minimise cognitive decline, continued research is needed to better understand the action of NIBS techniques and the influence of individual differences if we are to fully harness their value.

Therein, this collection of articles represents cross-modality research at the cutting-edge of advances in the multifaceted domain of visual attention-related processing. Crucially, each contribution in this Special Issue attests to the ageing process being dynamic and not simply characterised by global deficits. Similar inferences can be made in the context of various forms of cognitive impairment (MCI, MCI-LB, AD), where elements of preserved function have also been identified in the articles presented here. These considerations are highly important as they have direct consequences for prospective interventions, with the presence of individual differences implying that a one-size fits all approach will fail to incorporate the complexities outlined in the showcased research. By addressing the integrity of function via a unique array of approaches, encompassing novel methodologies and innovative concepts, these studies ultimately culminate in improving (1) our understanding of age-related changes, (2) the identification of potential signatures of a range of disease processes, and (3) interventions to attenuate the impact of observed decline and maintain quality of life. With a firm emphasis on the development of naturalistic applications, which test the assumptions put forward by basic research, an increase in the type of investigations advocated here will ensure the real-life relevance of future endeavours.

Conflicts of Interest: The authors declare no conflict of interest.

References

1. Maniglia, M.R.; Souza, A.S. Age Differences in the Efficiency of Filtering and Ignoring Distraction in Visual Working Memory. *Brain Sci.* **2020**, *10*, 556. [CrossRef] [PubMed]
2. Henderson, S.E.; Lockhart, H.A.; Davis, E.E.; Emrich, S.M.; Campbell, K.L. Reduced Attentional Control in Older Adults Leads to Deficits in Flexible Prioritization of Visual Working Memory. *Brain Sci.* **2020**, *10*, 542. [CrossRef] [PubMed]
3. Revie, L.; Hamilton, C.A.; Ciafone, J.; Donaghy, P.C.; Thomas, A.; Metzler-Baddeley, C. Visuo-Perceptual and Decision-Making Contributions to Visual Hallucinations in Mild Cognitive Impairment in Lewy Body Disease: Insights from a Drift Diffusion Analysis. *Brain Sci.* **2020**, *10*, 540. [CrossRef] [PubMed]
4. Barel, E.; Tzischinsky, O. The Relation between Sustained Attention and Incidental and Intentional Object-Location Memory. *Brain Sci.* **2020**, *10*, 145. [CrossRef] [PubMed]
5. Polden, M.; Wilcockson, T.D.; Crawford, T.J. The disengagement of visual attention: An eye-tracking study of cognitive impairment, ethnicity and age. *Brain Sci.* **2020**, *10*, 461. [CrossRef] [PubMed]
6. Torrens-Burton, A.; Hanley, C.J.; Wood, R.; Basoudan, N.; Norris, J.E.; Richards, E.; Tales, A. Lacking pace but not precision: Age-related information processing changes in response to a dynamic attentional control task. *Brain Sci.* **2020**, *10*, 390. [CrossRef] [PubMed]
7. Huizeling, E.; Wang, H.; Holland, C.; Kessler, K. Age-related changes in attentional refocusing during simulated driving. *Brain Sci.* **2020**, *10*, 530. [CrossRef] [PubMed]
8. Hanley, C.J.; Alderman, S.L.; Clemence, E. Optimising cognitive enhancement: Systematic assessment of the effects of tdcs duration in older adults. *Brain Sci.* **2020**, *10*, 304. [CrossRef] [PubMed]

Article

Age Differences in the Efficiency of Filtering and Ignoring Distraction in Visual Working Memory

Mariana R. Maniglia [1] and Alessandra S. Souza [2,*]

[1] Department of Psychology, Ribeirão Preto School of Philosophy, Science and Literature (FFCLRP), University of São Paulo, Avenida Bandeirantes 3900, Ribeirão Preto/SP 14040-901, Brazil; marianamaniglia@usp.br

[2] Department of Psychology, Cognitive Psychology Unit, University of Zurich, Binzmühlestrasse 14/22, 8050 Zurich, Switzerland

* Correspondence: a.souza@psychologie.uzh.ch

Received: 27 June 2020; Accepted: 8 August 2020; Published: 14 August 2020

Abstract: Healthy aging is associated with decline in the ability to maintain visual information in working memory (WM). We examined whether this decline can be explained by decreases in the ability to filter distraction during encoding or to ignore distraction during memory maintenance. Distraction consisted of irrelevant objects (Exp. 1) or irrelevant features of an object (Exp. 2). In Experiment 1, participants completed a spatial WM task requiring remembering locations on a grid. During encoding or during maintenance, irrelevant distractor positions were presented. In Experiment 2, participants encoded either single-feature (colors or orientations) or multifeature objects (colored triangles) and later reproduced one of these features using a continuous scale. In multifeature blocks, a precue appeared before encoding or a retrocue appeared during memory maintenance indicating with 100% certainty to the to-be-tested feature, thereby enabling filtering and ignoring of the irrelevant (not-cued) feature, respectively. There were no age-related deficits in the efficiency of filtering and ignoring distractor objects (Exp. 1) and of filtering irrelevant features (Exp. 2). Both younger and older adults could not ignore irrelevant features when cued with a retrocue. Overall, our results provide no evidence for an aging deficit in using attention to manage visual WM.

Keywords: attention; working memory; aging; filtering; ignoring; precue; retrocue

---

1. Introduction

It is widely recognized that healthy older adults show a decline in working memory (WM) compared to younger adults [1–3]. WM is a limited-capacity system that maintains a small amount of information available in mind for ongoing cognitive processing. One prominent explanation of WM age-related decline states that older adults have difficulties in using attention to efficiently manage the contents of WM. For example, one may use attention to filter out relevant from irrelevant information, thereby selecting only the relevant information to be encoded to WM [4]. Attention could also be used to ignore irrelevant information (i.e., distractors) presented while WM is engaged in maintaining other information [5,6]. The abilities to filter distractors at encoding (hereafter simply "*filtering*") and to ignore distractors during memory maintenance (hereafter "*ignoring*") have been shown to bear different relations to an individual's WM capacity [7,8]. Furthermore, these abilities can be directed to different objects [9] or different features of an object [10,11].

In the present study, we investigated the hypothesis that age-related cognitive decline in visual WM can be explained by decrease in the ability to filter and ignore distraction. We assessed this hypothesis across two experiments that required participants to attend to different visual objects (Exp. 1) or different features of an object (Exp. 2). In the following sections, we will review the evidence for age differences in these abilities and contextualize our research questions.

1.1. Attending to Different Objects

Our visual surroundings are full of visual items: people, animals, and objects. At any given moment, there is too much information for our cognitive system to process. Attention allows us to focus on a subset of these items thereby increasing the processing of relevant and decreasing processing of irrelevant information. For example, if we are looking for our keys, we will ignore the cat and the children running around. If we do not, we might lose focus and forget the places we have already searched.

Attention serves to control which information gets access to WM (for a review see [12]), and studies have shown that the ability to filter and ignore irrelevant information contributes to WM capacity [7,8,13]. Now the question stands: Do deficits in the ability to filter and ignore irrelevant information explain age-related decline in WM? To answer this question, McNab and colleagues [9] used smartphones to assess a sample of over 29,000 people in a game-like task requiring filtering and ignoring of distracting information. In this study, the task consisted in remembering the positions of red circles that appeared on a 4×4 grid for 1 s, maintaining them over another 1-s delay, and finally recalling all of the positions by clicking on the respective locations on the grid. Distracting information (yellow circles) could be presented together with the red circles for study thereby requiring the filtering of this information, or they appeared during the delay period, in which case they had to be ignored. Older adults showed larger declines in performance in the presence of distraction than younger adults, but this distraction effect was substantially larger when it occurred during the delay period (thereby requiring ignoring) compared to encoding (requiring filtering). McNab et al. proposed that the greater age-related impairing in ignoring may be explained by the distractors being encoded to WM in the ignore condition by both age groups; however, older adults may have difficulties judging the temporal order of memoranda vs. distractors leading to confusions at recall. There are other alternative explanations though. For example, older adults may require longer postdistraction delays than younger adults to resolve interference due to the slower processing speed associated with aging [14].

In our study, we included conditions to distinguish between these explanations: (a) a condition in which the distractors had to be encoded and maintained in WM instead of ignored (i.e., a high-load baseline) and (b) a condition in which more time was provided to ignore the distractors (ignore+delay condition). The inclusion of a high-load baseline allows one to measure the degree in which performance in the ignore condition reflects encoding of the distractors: if this is the case, performance in the ignore condition should be similar to the high-load baseline. If the distractors are only partially encoded to WM, then performance in the ignore condition should be somewhat better than the high-load baseline. Alternatively, the ignore condition could be even associated with worse performance than the High-Load condition, due to the increased demand of trying to remove distractors and not succeeding leading to loss of relevant items. The inclusion of the ignore+delay condition can clarify whether the difference between the filtering and ignoring conditions is explained by the delayed time-course of distraction exclusion in older age. In the study of McNab et al. [9], there was a 2-s delay between distractor and test in the filtering condition, whereas this time-separation was of only a 1-s in the ignore condition. Several studies have shown that older adults show delayed onset of Event-Related Potential (ERP) markers of distractor-exclusion in filtering tasks compared to younger adults, even in conditions in which equal behavioral performance was obtained between age groups [15,16]. If older adults are slower than younger adults to remove distractors from memory and this explains their worse performance in the ignore condition, then the inclusion of an ignore+delay condition should abolish the age-related decline observed in ignoring.

In sum, older adults seem to have more difficulty ignoring than filtering information, but further research is needed to clarify the reasons for this dissociation. The goal of Experiment 1 was to address the possibilities that this is due to the encoding of the distractors or due to the faulty removal of the distractors from visual WM.

1.2. Attending to Different Features

The ability to filter or ignore distractors is assumed to serve visual WM not only when attention is directed to different objects but also when one has to selectively attend to only some of the features of a single object. For example, one may see a large red star and a small blue circle. Remembering the conjunction of multiple features places larger demands on visual WM than when simpler (single-feature) objects are maintained [17–19], and this ability is also subjected to large age-related cognitive decline [20,21]. How filtering and ignoring are used when irrelevant information steams from the same (attended) object has been less investigated and, to the best of our knowledge, no study has investigated how it changes in aging.

Research with younger adults have provided mixed answers regarding the question of whether people are able to filter irrelevant features of an object. This question has been examined by comparing conditions in which participants store single-feature objects vs. multifeature objects when (a) only one of the features of the multifeature object is relevant (filtering condition) or (b) all features of the objects are relevant. One of the first studies to address this issue was done by Luck and Vogel [22]. They observed that memory was not affected by the number of visual features within an object, suggesting that participants always encoded all features to visual WM irrespectively of their relevance, and storing more features was not detrimental. More recent studies have provided evidence contradicting these results, showing that storing multiple features consumes more WM capacity [11,17,19,23]. Whether people can filter irrelevant features when they have foreknowledge of which features are relevant has yielded mixed results. Some studies [11,24,25] observed that younger participants could not voluntarily filter irrelevant features, however, other studies have found nearly perfect filtering [26–28].

These studies pertain to filtering of irrelevant features at encoding. Information regarding the relevance of some features over others may become available, however, only after the objects were already encoded to WM. A handful of recent studies have also investigated whether people can selectively weight features already encoded to WM, or in other words, whether they can "ignore" irrelevant features in WM [10,29,30]. In these studies, participants encoded a set of multifeatured objects (e.g., colored and oriented gratings). During the retention interval, a cue indicated which feature was relevant for the upcoming test (either color or orientation) or one of the features was tested with no prior warning (no-cue baseline). Compared to the no-cue baseline, cueing the relevant feature during the retention interval yielded better performance.

As reviewed above, only a handful of studies have investigated the ability to filter and ignore features, and these studies were done with younger adults. To the best of our knowledge, no study investigated these abilities in aging. The goal of Experiment 2, therefore, was to assess age-related decline in filtering and ignoring irrelevant features.

1.3. The Present Study

Our research questions and hypotheses were preregistered (https://osf.io/jmhwv). As stated in the preregistration, we aimed to investigate age differences in the ability to filter and ignore information in visual WM. In Experiment 1, we investigated the ability to filter and ignore distractor objects. We followed-up on the study of McNab et al. [9], which showed evidence for the greater age-related decline on the ability to ignore distractor objects during maintenance compared to filtering distractors at encoding. In Experiment 1, we tested two hypotheses regarding this age deficit in ignoring. The first possibility is that the sudden onset of the distractors during the retention interval grabs the attention of older adults, leading them to fully or partially encode the distractors to WM. Younger adults, however, can fully ignore the distractors. The second possibility is that both younger and older adults encode the distractors to WM, but younger adults are faster to remove the distractors from WM compared to the older adults. We addressed these possibilities by including conditions that require people to fully encode the distractors to WM (high-load baselines) and conditions that vary the available time to remove distractors from WM (ignore-delay condition).

Experiment 2 extended the investigation of filtering and ignoring abilities to situations pertaining to the encoding and maintenance of multifeature objects. The investigation of the selective weighting of features stored in visual WM is quite recent. Only a few studies have assessed this ability and only among younger adults. Hence, it is not clear whether the ability to ignore irrelevant features would be subjected to more age-related decline than the ability to filter features, similar to the observed for filtering and ignoring of whole objects.

In summary, the current study allows us to investigate four main research issues concerning age-differences in visual WM: (1) Is the inability to ignore distractions in older adults due to their complete or partial encoding? (2) Do older adults take longer to disengage from the information they should ignore? (3) Is there an age difference in the ability to filter irrelevant features in visual WM? and (4) Is there an age difference in the ability to ignore irrelevant features in visual WM?

2. Materials and Methods

2.1. Participants

Thirty younger students (19–32 years old, $M = 25.7$; 17 women) and thirty older adults (65–79 years old; $M = 72.6$; 14 women) participated in the study. Younger adults were students from the University of Zurich and older adults were Zurich community-dwelling individuals. Participation was compensated with CHF 15 per h or extra course credit (in case of students), and the experimental session lasted between 2 and 3 h. All participants signed the informed consent at the beginning of the experiment. The study protocol was in line with the ethical guidelines of the Institutional Review Board of the University of Zurich and did not require special approval.

All participants confirmed that they have normal or corrected-to-normal visual acuity and color vision. Older adults were assessed with the Mini-Mental Status Exam (MMSE) and all scored higher than 28 ($M = 29.3$).

2.2. Materials and Apparatus

Participants were tested in a group laboratory containing four desks, each with a computer. Up to four people could be tested simultaneously. The experimental session was composed of four tasks. The tasks were programmed in MATLAB using the Psychophysics toolbox [31,32]. The experimental tasks proceeded in the following order. First, participants completed two perceptual control tasks measuring their ability to match the color and orientation of visual objects. Next, participants completed the task for Experiment 1 (filter and ignore objects). At the end of this experiment, participants were allowed a 10-min break. Finally, participants completed the task of Experiment 2 (filter and ignore features).

2.3. Procedure

2.3.1. Perceptual Matching Tasks

These tasks assessed the motor and perceptual abilities of the participants, and they provided older adults the possibility to practice the continuous adjustment scale used in the memory tasks.

In the color version of the task, two colored dots appeared side-by-side against a grey background (RGB 128 128 128). The dot on the left was colored and served as a target. The dot on the right was dark grey (RGB 90 90 90) and served as the probe. Both dots were surrounded by a color wheel, which was defined in the CIELAB color space with $L = 70$, $a = 20$, $b = 38$, and radius = 60 [33]. The task was to adjust the color of the probe dot to match, as precisely as possible, the color of the target dot. Participants adjusted the probe's color by moving the mouse around the color wheel, which caused the color of the probe dot to change to the same color as currently selected by the mouse. A mouse-click served to confirm a response, upon which a feedback screen was showed indicating the response of the participant and the correct response.

In the orientation version of the task, the target and the probe were white isosceles triangles presented against a grey background (RGB 128 128 128). Participants had to adjust the orientation of the probe triangle (on the right) to match the orientation of the target triangle (on the left) by moving the mouse around. The tip of the triangle pointed in the same direction as the mouse cursor. As for the other task, a mouse click locked the response and was followed by the presentation of a feedback screen.

Participants completed 4 practice trials and 50 test trials in each task. In both tasks, stimuli were randomly selected from the 360 values of the color wheel or of orientation with the constrain that the selected values for each trial were at least 5° apart from each other. The dependent variable in these tasks was the distance between the reported feature value and the true feature value of the target.

2.3.2. Experiment 1

The experimental task was modeled after the one used by McNab et al. [9]: the position of colored dots on a 10 × 10 grid had to be remembered (see example in Figure 1). The grid was shown in white (RGB 255 255 255) against a uniform grey background (RGB 128 128 128). Participants were told to remember the positions of blue (RGB 0 0 255) and green (RGB 0 255 0) dots, and to ignore any red (RGB 255 0 0) dots that appeared. At test, participants were asked, first, to reconstruct the positions of the blue dots by clicking on the corresponding cells of the grid, followed by reconstruction of the position of the green dots (in case these were presented). The recall of the different types of dots was indicated by the color of the outer frame of the grid. Participants had to click on as many dots as presented in that color before the next recall prompt or the end of the trial was presented. There was no time limit for the recall.

The task was completed across two blocks. In the baseline block, participants were instructed to remember the positions of only blue dots (Low-Load condition) or blue and green dots (High-Load conditions) as detailed below. The number of blue dots was seven for younger and five for older adults. The number of green dots presented in the High-Load conditions was always three for both groups. Different levels of load were presented for each age group because previous research [9] indicated that older adults can store, on average, one-and-half objects less than younger adults do. By reducing the memory load for older adults, we hoped to test the filtering and ignoring abilities at similar levels of difficulty for both age groups. In all conditions, blue dots were tested first, and in the case of the High-Load conditions, this was followed by the test of the green dots. This is because our main question pertains to how well the blue dots can be maintained across all experimental conditions. There were three types of conditions in this block. In the *Low-Load condition* (Figure 1A), only blue dots were showed on the screen for 1000 ms, followed by 1000 ms retention and the recall test. In the *High-Simultaneous Load condition* (Figure 1B), blue and green dots were presented simultaneously on screen for 1000 ms, followed by 1000 ms retention and, lastly, by the recall tests. In the *High-Sequential Load condition* (Figure 1C), blue dots were shown for 1000 ms, followed by the presentation of the green dots for 1000 ms, and then the recall tests. These three conditions serve as the low and high memory-load baseline conditions that allowed us to assess how well participants can filter and ignore distracting information in the main experimental conditions.

In the experimental block, participants were instructed to remember the positions of the blue dots and ignore the red dots. This block again composed of three conditions. In the *Filter condition* (Figure 1D), blue and red dots were presented simultaneously on the screen for 1000 ms, followed by 1000 ms retention, and then the recall test. In the *Ignore condition* (Figure 1E), blue dots were presented first for 1000 ms, followed by the red dots for 1000 ms. The recall test was presented immediately after the offset of the red dots. In *Ignore+Delay condition* (Figure 1F), the recall test was presented after an additional delay of 1000 ms. The inclusion of the postdistraction delay allows us to test for the possibility that older adults need more time to remove the red dots from their WM.

Participants completed 120 trials, divided into the two blocks (whose order was counterbalanced across participants). Given that there were six experimental conditions (i.e., Low-Load, High-Simultaneous Load, High-Sequential Load, Filter, Ignore, and Ignore+Delay), there were

20 trials per design cell. Trials of each condition were randomly intermixed within each block. Six practice trials before each block were done. Opportunities for short breaks were provided every 10 trials.

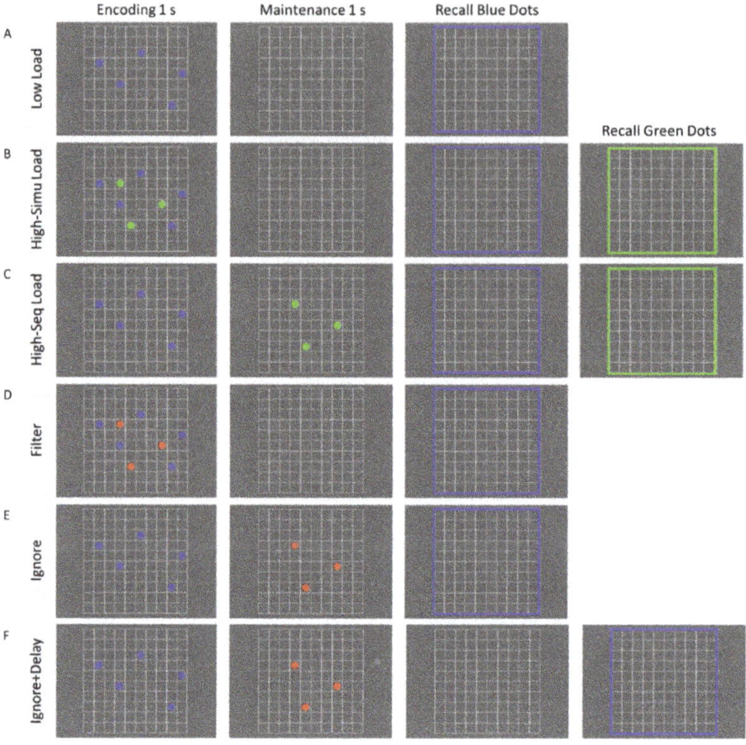

Figure 1. Example of conditions in Experiment 1. In the *Low-Load condition* (**A**) participants encoded blue dots and later reproduced them by clicking on the corresponding grid cells. In *High-Simultaneous Load* (**B**) and *High-Sequential Load* (**C**) conditions, participants encoded both blue and green dots. In the *Filter condition* (**D**), participants encoded the blue dots while ignoring the red dots presented at encoding. In the *Ignore condition* (**E**), participants encoded first the blue dots and ignored the red dots presented during the maintenance phase. In the *Ignore+Delay condition* (**F**) the recall test was presented after an additional delay of 1000 ms.

2.3.3. Experiment 2

The task in Experiment 2 consisted of remembering visual objects that vary in one or two continuous feature dimensions (i.e., color and orientation), and at test they reproduced one of these dimensions on a continuous scale (see Figure 2).

In the *Single-Feature* baseline conditions, the memory array consisted either of colored dots (which do not contain orientation information; see Figure 2A) or white triangles (which do not contain color information; see Figure 2B). In the test, participants reproduced the single feature of one of the objects in the memory array. In the *Dual-Feature* baseline condition (Figure 2C), participants were presented with colored isosceles triangles, and they remembered both the color and the orientation of these objects. In the test, one object was randomly selected and participants reproduced either its color or its orientation (with equal probability). In the *Filter condition* (see Figure 2D), a precue was presented before the onset of the memory array and indicated with 100% validity whether the color or the orientation of one of the objects would be tested at the end of the trial. The precue allowed participants to only encode the relevant feature to WM. In the *Ignore condition* (see Figure 2E), a retrocue was presented during the retention interval indicating with 100% certainty whether color or orientation would be tested later in the trial. The retrocue allowed participants to ignore, at the test, the uncued feature.

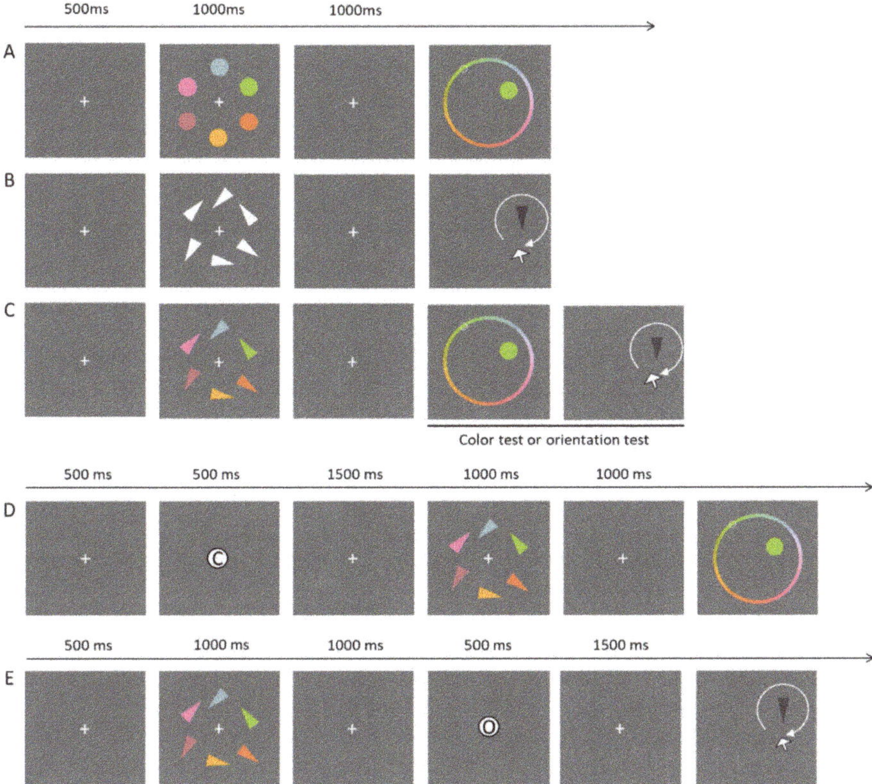

Figure 2. Example of conditions in Experiment 2. In *Single-Feature baseline* conditions, participants encoded only colors (**A**) or the orientation of the triangles (**B**) and then they indicated the color of the target by clicking on the color wheel or the correct triangle's orientation by moving the mouse to rotate the probe. In *Dual-Feature baseline* (**C**), participants encoded colored triangles, and then either the color or the orientation of an item was probed. In *Filter condition* (**D**), a cue was shown in the middle of the screen indicating the to-be-tested feature (C = color; O = orientation) before the onset of the memory array, allowing participants to only encode the relevant feature to WM. In *Ignore condition* (**E**), a cue was shown during the retention interval, thereby allowing participants to ignore one of the features already stored in WM.

In all conditions, the trial started with a white fixation cross presented against a grey background (RGB 128 128 128) for 500 ms. Next, the memory array (colored dots, white oriented triangles, or colored triangles) was presented for 1000 ms. The memoranda were evenly spaced on an imaginary circle centered on the middle of the screen. Colors were randomly selected from 360 values distributed along with the same color wheel as defined for the color perceptual task. Orientations were selected randomly from 360° angles. After a brief retention interval (1000 ms), memory for one of the stimuli was tested by presenting a dark-grey stimulus (probe) at the location of one of the memoranda. If color was the tested feature, a color wheel was shown. Participants had to move the mouse along the color wheel to adjust the color of the probe to the color of the memory stimulus presented in that location.

If the orientation had to be reproduced, only the probe stimulus and the mouse cursor were presented. Participants had to move the mouse around to rotate the probe triangle to match the orientation of the memory triangle that was presented in the same location. Participants pressed the left mouse button to confirm their response. There was no time limit for the recall.

In precue trials (*Filter condition*), a letter (F for color—which is "Farbe" in German and R for orientation—"Richtung" in German) was presented in the middle of the screen for 500 ms indicating which feature (color or orientation) was relevant for the test. The memory array was presented 1500 ms after the offset of the precue. In retrocue trials (*Ignore condition*), the letter F or R was presented 1000 ms after the offset of the memory array. The cue was on screen for 500 ms, and 1500 ms thereafter, the test-array appeared.

The number of objects in the memory array was 6 for younger adults and 4 for older adults. Again, our goal with the presentation of different memory loads between age groups was to present tasks of equivalent difficulty for these groups. Experiment 2 consisted of 400 trials divided into 2 blocks of 200 trials each: a baseline block consisting of the Single- vs. Dual-Feature trials; and an experimental block consisting of the Filter and Ignore trials. There were 4 training trials before each block. Within each block, the type of trial (Dual-Feature vs. Single-Feature; or precue vs. retrocue) was randomized with the constrain that each design cell had an equal number of trials (note that the dual-feature condition had 100 trials, with 50 trials ending in a color test and 50 trials in an orientation test). The order of blocks was counterbalanced across participants. A short break was provided between blocks.

3. Data Analysis

The practice trials were excluded from all analyses. In Experiment 1, the dependent variable was the proportion of correctly recalled blue dots across conditions. In Experiment 2, our main dependent variable was the error in reporting the cued feature (either color or orientation) of the tested item across conditions. In the perceptual matching tasks, the dependent variable was the error in matching the feature of the perceptually visible target item. Recall error is the absolute value of the deviation between the true feature value of the tested item and the participants reported feature value.

We analyzed the data using two approaches: *Bayesian model comparison* and *Bayesian parameter estimation*. The first approach compares the likelihood of a stated model given the data. This analysis yields an odds ratio, denominated Bayes factor (BF_{10}), which quantifies the evidence for supporting the stated model against the null model. Evidence for the null model over the stated model can be obtained by computing $1/BF_{10}$ (BF_{01}) [34,35]. By comparing models that include different predictors, this procedure allows gathering evidence for or against the presence of main effects and interactions between predictors. Here, we used Bayesian ANOVAs (hereafter BANOVA) and Bayesian *t*-tests to gather evidence for and against our hypotheses.

The second approach computes the relative credibility of all candidate parameter values that can describe the data given the specified model, which yields distributions of the parameter values known as the posterior. To assess the credibility of the parameter, we consider the interval that covers 95% of its posterior (hereafter, the highest density interval–HDI). If the HDI of a parameter does not include 0, or if it is outside of a region around the null, namely, a Region of Practical Equivalence (ROPE; which here was defined around an effect size between −0.1 and 0.1), then the estimated value is considered credibly different from 0. Both analyses were implemented in R [36]: BANOVAs and *t*-tests used the BayesFactor package [37], and parameter estimation used the BEST functions [38].

In our preregistration, we indicated that our main interest in both experiments was in comparing performance in the Filtering and Ignoring conditions concerning both the Low-Load and their respective matched High-Load conditions (High-Simultaneous and High-Sequential, respectively) and to assess age-related change on these. We proposed to compare them using a relative scoring that would combine both comparisons (e.g., (Filtering–High-Simultaneous)/(Low–High-Simultaneous)). However, given the performance levels, we observed in the experiments, this computation yielded uninterpretable values. This happened because some scores in the Filtering and Ignoring conditions were even better than in the Low-Load condition (In our preregistration, we only considered that performance in these distractor conditions would be in-between the Low-Load condition and the High-Load condition. It turns out that performance tended to be even better in these conditions than in the Low-Load baseline). Therefore, we decided to report separate comparisons with each of these load baselines to allow for proper testing of our stated hypotheses. For these comparisons, we used a two-tailed *t*-test. For comparisons between age-groups, however, we used a one-tailed *t*-test to more specifically test the hypothesis that older adults perform more poorly than younger adults.

4. Results

The study materials, data, and analysis scripts are available at the Open Science Framework at https://osf.io/ezr3m/.

4.1. Experiment 1

For one participant in the younger group, data of 10 trials was lost due to a computer crash.

Figure 3 shows performance in Experiment 1 as a function of the experimental condition for the two age groups. Overall, younger adults performed better in the task than older adults, even though we provided younger participants with a higher memory load (7 blue items) than older adults (5 blue items). Figure 3 also shows that there is no sign of interactions between experimental condition and age. Accordingly, a BANOVA including age (younger vs. old) and condition (Low, High-Simu, High-Seq, Filter Ignore, and Ignore+Delay) showed that the best model of this data included only the main effects of condition and age ($BF_{10} = 2.90 \times 10^{58}$). This model was substantially preferred to the model including the interaction term ($BF_{10} = 0.27$). There was substantial evidence to include the main effects of age ($BF_{10} = 3.56 \times 10^{53}$) and condition ($BF_{10} = 6.93 \times 10^{4}$) in the best model. We also assessed age effect only on the conditions involving distraction (Filter, Ignore, and Ignore+Delay). The best model of this data only included the main effects of age and condition ($BF_{10} = 2.96 \times 10^{9}$), and there was substantial evidence to exclude the interaction term ($BF_{10} = 0.17$). These results indicate that although older adults were able to recall fewer positions correctly, the performance of older and younger adults was similarly affected by memory load and distraction presence.

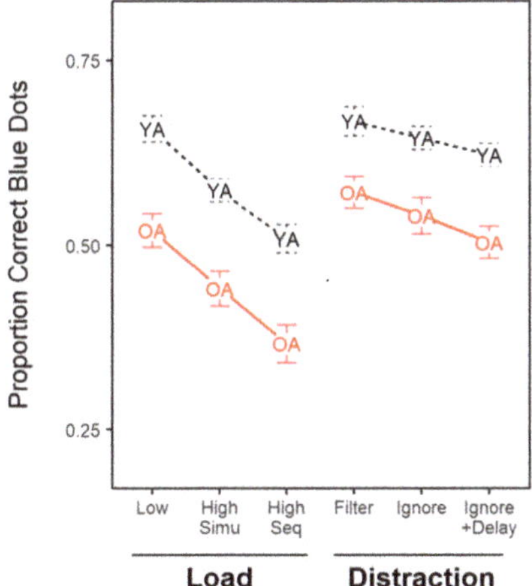

Figure 3. Proportion of correctly recalled blue dots in Experiment 1 for each age group (YA = younger adults, OA = older adults) across the experimental conditions that varied in memory load (Low, High-Simultaneous, and High-Sequential) and distractor presence (Filter, Ignore, and Ignore+Delay).

4.1.1. Filtering Ability

Table 1 shows contrasts between the *Filter* condition and the *Low-* and *High-Load* conditions, respectively. As indicated in Table 1, the *Filter* condition tended to yield better performance than the *Low-Load* condition, and this improvement was credible for older adults. This result is the opposite of what is predicted by a deficit in filtering irrelevant objects. We tested for the directional hypothesis that this filtering score was lower for the older adults than for the younger adults with a one-sided *t*-test, which our results show clear evidence against.

The comparison against the *High-Simultaneous* condition indicated that the *Filter* condition produced better performance in both age groups. We tested for an age difference in filtering against this baseline assuming that older adults performed worse than younger, but our results showed strong evidence against this difference. If anything, the older adults had a larger filtering benefit than the younger adults.

4.1.2. Ignoring Ability

Table 1 also presents the contrast between the *Ignore* condition and the *Low-* and *High-Load* conditions, respectively. The contrast of the *Ignore* condition to the *Low-Load* condition indicated the absence of credible changes in performance for both younger and older adults. Critically, there was substantial evidence against age differences, indicating that older adults were as able as younger adults to ignore irrelevant objects.

Comparison to the *High-Sequential Load* condition revealed better performance in the *Ignore* condition for both older and younger adults. This outcome indicates that the older adults did not encode the items that they should ignore; indeed, encoding additional items overloads WM capacity leading to worse performance in the *High-Sequential* condition, which was not the case for the *Ignore* condition. Again, there was no evidence for an age-related difference in ignoring ability.

4.1.3. Delay Effect: More Time to Ignore Distractors?

Our last question was whether providing a delay after the presentation of the distractors could improve the ignoring of the distractors. Our data showed, unexpectedly, that neither younger nor older adults had problems ignoring the distractors as presented in the previous section. Indeed, as shown in Table 1, the presentation of a longer delay after the distractors was associated with worse performance in the *Ignore+Delay* condition compared to the *Ignore* condition. There was no age-related difference in this effect.

4.2. Perceptual Matching Tasks

Before Experiment 2, participants completed two perceptual matching tasks that allowed us to assess their motor and perceptual abilities with regards to the reproduction of colors and orientations. Younger adults (M = 2.80, SD = 0.85) tended to reproduced the colors with smaller error compared to older adults (M = 3.19, SD = 1.33), but this difference was not credible, BF_{10} = 0.5, with 14% of the credible effect size (M = −0.32) values within the ROPE (i.e., effect size between −0.1 and 0.1). For orientation reproduction, younger adults (M = 3.90, SD = 2.70) showed similar performance to older adults (M = 3.57, SD = 1.64), and for this comparison, the null hypothesis was preferred by a factor of about 3, BF_{10} = 0.30, with evidence of 25% of the credible effect size values (M = 0.06) within the ROPE. These results indicate that any differences in performance between younger and older adults in terms of memory performance in Experiment 2 cannot be attributed to worse perceptual-motor abilities.

Table 1. Analysis of the data of Experiment 1. Mean recall error and 95% highest density interval (HDI) estimated from Bayesian *t*-tests assessing (1) the age effect in each condition, (2) the filter, ignore, and delay effect within each age group, and (3) the age effect on the size of the filter, ignore, and delay effects.

	Age Group				*Age Effect (Older < Younger)*					
	Younger		**Older**		**Raw Score**		**Effect Size**			
Condition	M	95% HDI	M	95% HDI	M	95% HDI	M	95% HDI	ROPE	BF_{10}
Low	0.66	[0.62,0.70]	0.52	[0.49,0.56]	0.14	[0.09,0.19]	1.47	[0.85,2.10]	0	5×10^4
High-Simu	0.57	[0.54,0.61]	0.44	[0.40,0.48]	0.13	[0.08,0.18]	1.38	[0.82,2.05]	0	3.4×10^4
High-Seq	0.51	[0.48,0.54]	0.36	[0.33,0.40]	0.15	[0.10,0.19]	1.57	[0.92,2.28]	0	1.4×10^5
Filter	0.67	[0.63,0.71]	0.58	[0.54,0.61]	0.10	[0.05,0.15]	1.03	[0.45,1.60]	0	162
Ignore	0.64	[0.61,0.68]	0.54	[0.51,0.58]	0.10	[0.06,0.15]	1.10	[0.55,1.73]	0	970
Ignore+Delay	0.62	[0.59,0.66]	0.50	[0.47,0.54]	0.12	[0.07,0.17]	1.33	[0.71,1.89]	0	4500
Contrasts										
	Filter = Low-Load									
Raw score	0.01	[−0.02,0.04]	0.05	[0.02,0.08]	−0.04	[−0.08,0.0]	−0.50	[−1.07,0.0]	0.05	0.09
Effect size	0.14	[−0.24,0.53]	0.70	[0.28,1.12]						
p(ROPE)	0.31		0							
BF_{10}	0.25		57							
	Filter = High-Simu									
Raw score	0.09	[0.07,0.12]	0.13	[0.10,0.17]	−0.04	[−0.08,0.0]	−0.48	[−1.04,0.06]	0.07	0.10
Effect size	1.52	[0.96,2.09]	1.38	[0.85,2.09]						
p(ROPE)	0		0							
BF_{10}	3.9×10^6		5.3×10^5							
	Ignore = Low-Load									
Raw score	−0.01	[−0.04,0.01]	0.02	[−0.01,0.05]	−0.03	[−0.07,0.0]	−0.41	[−0.96,0.11]	0.09	0.11
Effect size	−0.18	[−0.58,0.19]	0.22	[−0.14,0.62]						
p(ROPE)	0.25		0.20							
BF_{10}	0.32		0.39							
	Ignore = High-Seq									
Raw score	0.14	[0.11,0.17]	0.18	[0.13,0.22]	−0.04	[−0.09,0.01]	−0.44	[−0.95,0.13]	0.09	0.11
Effect size	1.83	[1.23,2.55]	1.66	[1.05,2.27]						
p(ROPE)	0		0							
BF_{10}	1.7×10^8		1.5×10^7							
	Ignore-Delay > Ignore									
Raw score	−0.02	[−0.04,0.00]	−0.04	[−0.06,−0.01]	0.01	[−0.02,0.05]	0.14	[−0.33,0.71]	0.23	0.16
Effect size	−0.41	[−0.82,−0.02]	−0.48	[−0.90,−0.10]						
p(ROPE)	0.05		0.02							
BF_{10}	0.07		0.06							

Note: Simu = Simultaneous; Seq = Sequential. For each effect, the evidence (BF) for the alternative hypothesis over the null is presented (BF_{10}) for a one-sided test. ROPE = probability of values within a region of practical equivalence (effect size between −0.1 and 0.1). For within-subject condition comparisons (e.g., Filter = Low-Load), we performed two-tailed *t*-tests. For between-subject comparisons (e.g., older < younger), we performed a one-tailed *t*-test.

4.3. Experiment 2

Figure 4 presents the recall error for reproducing the colors and orientations in Experiment 2. Overall, performance was best when participants encoded single-feature objects (colored dots or oriented triangles) and worse when they encoded dual-feature objects (colored triangles). The *Filter* conditions yielded performance similar to the *Single-Feature* conditions. The *Ignore* conditions, in contrast, yielded performance similar to the *Dual-Feature* conditions.

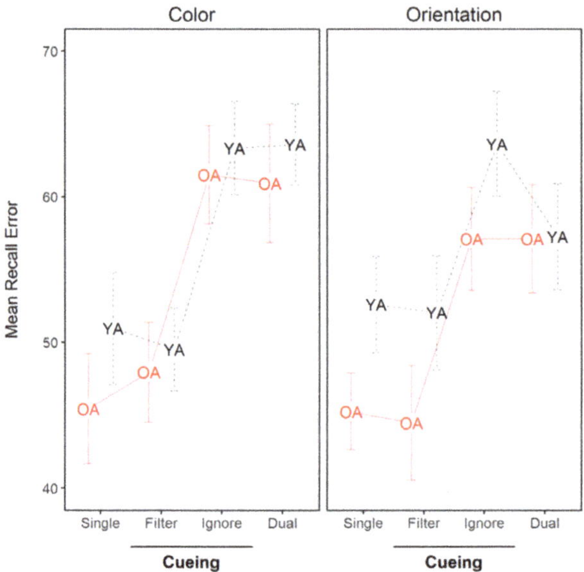

Figure 4. Mean recall error in Experiment 2 for each age group (YA = younger adults, OA = older adults), recall (color and orientation), and *baseline condition* (Single and Dual) and *cueing condition* (Filter and Ignore).

We ran a BANOVA on these data having conditions (Single, Filter, Ignore, Dual), recalled feature (color vs. orientation), and age group (younger vs. older) as fixed predictors, and participant as a random effect. The best model included only the main effect of condition ($BF_{10} = 1.42 \times 10^{31}$). The evidence against including the main effect of the recalled feature was ambiguous ($BF_{10} = 0.70$) as well as of including the main effect of age group ($BF_{10} = 0.63$). These results suggest that performance in Experiment 2 mainly varied with the condition.

To facilitate gathering evidence for age differences, we also ran two separate BANOVAs, one for recall of colors and one for recall of orientations. For color recall, the best model of the data only included the main effect of condition ($BF_{10} = 1.76 \times 10^{22}$). There was ambiguous evidence against excluding the main effect of age group ($BF_{10} = 0.48$). Critically, there was strong evidence against including the interaction of age group and condition ($BF_{10} = 0.10$). For orientation recall, the best model of the data also only included the main effect of condition ($BF_{10} = 9.99 \times 10^{11}$). There was ambiguous evidence against excluding the main effect of age group ($BF_{10} = 0.78$) and for excluding the age group × condition interaction ($BF_{10} = 0.74$). Notice that this is probably because the younger group tended to perform worse in the *Ignore* condition than the *Dual-Feature* condition.

As there was no credible main effect of recalled feature in the BANOVA (presented above), the analyses of the main effects of Bayesian Estimation (Table 2) were calculated from the average performance across recall of color and orientation.

Table 2. Analysis of recall error in Experiment 2. Mean recall error and 95% highest density interval (HDI) estimated from Bayesian *t*-tests assessing (1) the age effect in each condition, (2) the load (single/dual), filter, and ignore benefit in each age group, and (3) the age effect on the condition effect.

	Age Group				*Age Effect (Younger < Older)*					
	Younger		Older		Raw Score		Effect Size			
Condition	M	95% HDI	M	95% HDI	M	95% HDI	M	95% HDI	ROPE	BF_{10}
Single-Feature	51.8	[46.9,56.8]	45.0	[39.8,49.8]	6.84	[−0.15,13.7]	0.51	[−0.02,1.06]	0.05	0.10
Color	50.8	[45.4,56.2]	45.2	[39.7,50.3]	5.66	[−1.81,13.3]	0.40	[−0.14,0.93]	0.10	0.11
Orientation	52.8	[47.0,58.5]	44.9	[39.4,50.4]	7.92	[0.06,16.0]	0.52	[−0.02,1.10]	0.05	0.10
Dual-Feature	60.6	[55.7,65.5]	59.1	[54.8,63.5]	1.54	[−4.81,7.95]	0.07	[−0.39,0.67]	0.26	0.20
Color	63.7	[58.9,68.3]	60.9	[57.0,65.0]	2.73	[−3.31,8.83]	0.26	[−0.28,0.77]	0.20	0.15
Orientation	57.4	[51.4,63.6]	57.3	[51.2,63.2]	0.12	[−8.7,8.23]	−0.02	[−0.52,0.52]	0.30	0.26
Filter	50.7	[44.1,57.1]	45.6	[40.0,51.0]	5.06	[−3.79,13.2]	0.32	[−0.22,0.85]	0.15	0.14
Color	49.4	[43.4,55.4]	47.4	[42.0,52.5]	1.98	[−5.93,9.94]	0.14	[−0.41,0.66]	0.25	0.20
Orientation	52.1	[44.7,59.8]	44.0	[37.0,51.0]	8.07	[−2.02,18.5]	0.37	[−0.12,0.95]	0.09	0.12
Ignore	63.4	[58.6,68.6]	58.9	[54.6,63.6]	4.47	[−2.04,11.4]	0.35	[−0.19,0.89]	0.13	0.13
Color	63.2	[58.1,68.4]	61.0	[56.0,65.6]	2.22	[−5.11,9.1]	0.19	[−0.37,0.72]	0.23	0.18
Orientation	63.8	[57.9,70.0]	56.8	[51.2,62.6]	7.00	[−1.10,15.3]	0.45	[−0.09,1.01]	0.07	0.11
Contrasts										
Dual > Single										
Raw score	8.51	[4.91,12.1]	13.7	[9.45,18.0]	−5.15	[−10.7,0.42]	−0.47	[−1.04,0.03]	0.06	2.29
Effect size	0.90	[0.47,1.37]	1.21	[0.74,1.76]						
p(ROPE)	0		0							
BF_{10}	2402		1.6×10^5							
Single = Filter										
Raw score	1.2	[−2.93,5.17]	−0.80	[−4.09,2.55]	2.01	[−3.04,7.40]	0.19	[−0.33,0.73]	0.22	0.16
Effect size	0.11	[−0.27,0.49]	−0.11	[−0.48,0.28]						
p(ROPE)	0.34		0.36							
BF_{10}	0.22		0.22							
Dual = Filter										
Raw score	9.92	[6.40,12.10]	13.2	[8.93,17.40]	−3.20	[−8.57,2.24]	−0.32	[−0.86,0.21]	0.15	0.80
Effect size	1.08	[0.58,1.70]	1.17	[0.66,1.85]						
p(ROPE)	0		0							
BF_{10}	4776		1.7×10^5							

Table 2. *Cont.*

	Age Group				*Age Effect (Younger < Older)*					
	Younger		Older		Raw Score		Effect Size			
Condition	M	95% HDI	M	95% HDI	M	95% HDI	M	95% HDI	ROPE	BF_{10}
Single = Ignore										
Raw score	−11.4	[−14.8,−8.02]	−14.0	[−17.7,−10.6]	2.57	[−2.19,7.41]	0.33	[−0.24,0.84]	0.18	0.64
Effect size	−1.35	[−1.91,−0.84]	−1.49	[−2.17,−0.94]						
p(ROPE)	0		0							
BF_{10}	4.9×10^5		2.5×10^6							
Dual = Ignore										
Raw score	−3.14	[−6.34,0.17]	−0.22	[−3.54,3.03]	−2.88	[−7.57,1.72]	−0.32	[−0.86,0.20]	0.14	0.87
Effect size	−0.40	[−0.78,0.02]	−0.00	[−0.40,0.35]						
p(ROPE)	0.08		0.40							
BF_{10}	0.97		0.19							

Note: For each effect, the evidence (BF) for the alternative hypothesis over the null is presented (BF_{10}) for a one-sided test. ROPE = probability of values within a region of practical equivalence (effect size between −0.1 and 0.1). For most within-subject condition comparisons (e.g., Single = Filter), we performed two-tailed *t*-tests, except for the comparison of Dual > Single, for which a substantial body of research creates the expectation of worse performance in the Dual than the Single condition. For between-subject comparisons (e.g., younger < older), we performed a one-tailed *t*-test.

4.3.1. Filtering Ability

Table 2 presents the comparison of the *Filter* condition against the *Single-Feature* and *Dual-Feature* conditions. The contrast of *Filter* vs. *Single-Feature* provided evidence against credible differences for both age groups, and there was also substantial evidence against an age reduction in filtering efficiency. This is pattern of results expected given perfect filtering of the irrelevant feature at encoding.

The contrast of *Filter* vs. *Dual-Feature* showed smaller recall error in the *Filter* condition. There was substantial evidence that storing both features yielded worse recall error for both age groups (i.e., *Single* vs. *Dual* contrast). Older adults tended to show a larger cost in the Dual-Feature condition, but this effect was not credible.

Overall, the results are consistent with the hypothesis that feature filtering efficiency was unaffected by age.

4.3.2. Ignoring Ability

Table 2 presents the comparison of the *Ignore* condition against the *Single-Feature* and *Dual-Feature* conditions. The contrast *Ignore* vs. *Single-Feature* showed larger recall error in *Ignore* condition for both age groups, indicating that participants were unable to ignore the irrelevant feature in WM.

The contrast *Ignore* vs. *Dual-Feature condition* showed worse recall in the *Ignore condition*, particularly for the younger group. There was, however, no credible evidence for an age effect on ignoring ability.

5. General Discussion

The current study investigated age-related differences in the efficiency of using attention to manage the contents of WM, specifically in filtering and ignoring irrelevant objects (Exp. 1) and in filtering and ignoring irrelevant features of the object (Exp. 2).

Our experiments revealed the following main findings: older adults were as able as younger adults to efficiently filter and ignore distractor objects (Exp. 1). Furthermore, older and younger adults performed similarly in filtering distractor features (Exp. 2). This experiment also showed no evidence that younger and older adults could ignore a feature they already encoded to visual WM. Overall, Experiment 2 also pointed to a lack of evidence that older adults have deficits in managing the contents of visual WM in comparison to younger adults.

Next, we will discuss the implications of these findings in terms of attention to objects and features.

5.1. Attentional Modulation of Objects

McNab and colleagues [9] showed evidence that filtering of irrelevant objects would be relatively preserved in aging, but ignoring irrelevant objects would be impaired. Here, we were interested in investigating whether the inability to ignore distractions in older age was due to the full or partial encoding of the distractors. The results provide evidence of efficient ignoring of distractor objects in both age groups. Although we set up to test explanations of an ignoring deficit in aging [9], we failed to observe an age-related cognitive decline in ignoring as revealed by three findings: (1) there was no interaction between age and distractor condition, indicating that both age groups were similarly affected by the presence of distractors at encoding and during the delay; (2) comparison of performance in the Filter and Ignore conditions against our Low-Load baseline showed no evidence of a drop in performance, and (3) the better performance observed in the Filter and Ignore conditions compared to the High-Load baselines provided evidence against the hypothesis that distractors were encoded to WM.

Our results stand in opposition to the results from the study of McNab et al. [9]. There are several differences between our experiments that may explain these discrepant findings, which we will discuss below.

The first refers to the task setup. Our experiment involved a larger grid of spatial locations (10 × 10) than used by McNab et al. (4 × 4). We decided against a small grid because in such a grid, participants can use the strategy to remember the empty cells instead of the filled ones. Our decision to increase the grid was also related to the fact that in the study of McNab et al., younger adults were at ceiling performance in their baseline (no-distraction) condition, whereas this was not the case for the older adults. This potentially precluded them from properly measuring the impact of distraction in the younger group, because the impact of ignoring was assessed relative to this baseline.

A second difference between our studies refers to the memory load between age groups. Based on the McNab et al.'s data, we estimated that older adults could store, on average, 1.5 items less than younger adults, so we provided a load of 2-items less to the older adults in an attempt to keep task difficulty at a similar level between age groups. We note that this load difference cannot explain age differences in Experiment 1. On the one hand, although increasing memory load improves the chance of getting a response correctly, our grid had 100 locations, and hence chance probabilities were of 5% and 7% for older and younger adults, respectively. This gives younger adults an advantage of 2% over older adults; nonetheless, age differences between conditions were of, at least, 10% in average as shown in Table 1. On the other hand, we relied on a within-subjects comparison to determine the efficiency of filtering and ignoring, and for this comparison, guessing probability is constant. Despite this load manipulation, our results showed that older adults still performed more poorly overall than younger adults, suggesting that age differences might have been underestimated in McNab et al. due to the ceiling performance of younger adults. Critically, the memory load selected for the younger adults (7 items) was effective to bring performance of the younger adults below ceiling, permitting us to attest that the younger participants were performing at their maximal WM capacity even at our Low-Load (no-distraction) baseline.

A third difference refers to the fact that our Low-Load condition was randomly intermixed with the High-Load conditions. Arguably, the Low-Load (no-distraction) condition may produce even better performance if participants are not exposed to a situation in which they are constantly required to process the color of the dots in order to determine their category (blue vs. green). This requirement may have forced participants to remember not only the spatial locations of the dots, but also their color. Maintenance of color-location bindings is arguably more demanding than maintenance of only locations. This could have created a more difficult baseline than the one used by McNab et al., which did not involve this additional requirement. This may have led our study to underestimate the impact of distraction in the Filter and Ignore conditions. Future studies are needed to properly disentangle whether there are costs associated with Filtering and Ignoring when these conditions are contrasted to a baseline in which only location needs to be remembered vs. a baseline in which color-location bindings may need to be remembered. This will help to establish whether the costs of distraction are associated with the need to form bindings when distraction is presented. That said, we believe this does not undermine our main finding that age was not associated with impairments in ignoring information, because we still did not observe an interaction between age and distraction (Filter vs. Ignore), a contrast that is not confounded by baseline comparisons.

A fourth difference is that McNab et al. ran a span procedure, whereas our experiment consisted of a fixed memory load for several trials. In a span procedure, participants are exposed to trials with increasing memory load (in their study this ranged from 2 to 10 items). If the participant responds correctly in a number of trials (e.g., 2) with a given load, the load is increased in the next trial by one unit. When participants make a mistake, the trial is repeated, and if they repeatedly fail to recall all items correctly, the task ends. The last memory load in which perfect recall was observed is defined as the memory span of the participant. In such a procedure, memory span is measured within a couple of trials. This procedure, however, is probably disadvantageous for older adults because they are not provided enough trials to develop a proper strategy to manage the task. Younger adults might be faster in figuring out the best strategy to deal with the increasing memory load giving them an advantage in this type of procedure.

Overall, our study supports the conclusion that older adults are not impaired relative to younger adults in ignoring distractor objects when we provide them with a sufficiently large number of trials with the task and guarantee that younger and older adults were performing at maximal capacity.

In Experiment 1, we also wanted to determine whether older adults needed more time to ignore irrelevant objects. Aging is well-known to be associated with a decline in cognitive processing speed [39]. This age-related slowing also influences WM processes, e.g., the time required to encode a stimulus [40,41]. Contrary to our expectations, our attempt to provide more time to ignore the distractors in the *Ignore+Delay* condition did not help performance. Instead, performance decrements were observed indicating that there was time-based forgetting of the spatial representations of the to-be-remembered items due to the prolongation of the retention interval [42]. Importantly, this was observed for both age groups.

Regarding the ability to filter distractors, our findings replicate previous results that showed similar filtering performance between younger and older adults [9,16]. This suggests that older adults can maintain similar performance levels compared to their younger counterparts. Whether the mechanisms associated with this preserved performance are the same between age groups is still unclear. Some neurophysiological studies have observed age-related impairment in the degree of activity suppression in areas processing distractor information [40] and on the timing of onset of ERPs associated with suppression of distractors [16]. Together, these results suggest that older adults might need more time to efficiently deploy attention, but otherwise have intact attentional processing abilities, and hence, they can also manage their WM contents efficiently.

5.2. Attentional Modulation of Features

Experiment 2 examined the ability to selectively attend to only some of the features of a single object during encoding (also known as filtering) and during the maintenance period (also known as ignoring). Our results demonstrated similar performance in the *Filter* and *Single-Feature* conditions and strong evidence for better performance in the *Filter* than the *Dual-Feature* condition in both age groups. These results are in line with the hypothesis that participants used the precue to only encode the relevant feature to visual WM, efficiently filtering the irrelevant feature dimension. Encoding of this feature yielded a cost to performance as demonstrated by the *Dual-Feature* condition and replicating other studies showing WM capacity limitations in storing multifeature objects [17,18]. The evidence of a filtering feature benefit provides support for the initial encoding process that can be biased by feature-based selective attention [43,44]. Selectively encoding of relevant features can provide advantages to WM performance. Relevant features may be encoded faster and even more robustly in visual WM. This may also make the retrieval process less costly, considering that the more features are encoded, the more decision processes will be needed, increasing the probability of errors [44]. Critically, there was no age difference in filtering irrelevant features: younger and older adults were equally able to use the precue to select only the relevant feature for encoding into visual WM.

To determine whether it is possible to ignore a feature after it has been encoded into visual WM, we presented a retrocue during the retention interval indicating the relevant feature for the test. Performance in the *Ignore* condition was similar to the one in the *Dual-Feature* condition, indicating that the multifeature object was maintained intact until tested, and the irrelevant feature could not be ignored. If feature-based attention could modulate the selection of a single feature for recall, then we would expect similar performance in the *Ignore* and *Single-Feature* conditions or at least better performance than in the *Dual-Feature* condition, which our data clearly shows evidence against.

There are not many studies addressing the ability to ignore irrelevant features. Two previous studies compared a Dual-Feature condition to a feature retrocue condition and reported feature retrocue benefits [10,29]. It is unclear why we could not replicate these results here. We can rule out the possibility that the participants did not understand the meaning of the cue: the same cues were used in the *Filter* and the *Ignore* conditions, which were randomly intermixed across trials. Participants were extremely efficient in using the cue if it appeared before the memory array, thereby yielding filtering benefits, but not if it appeared after the memory array, which did not produce an ignoring effect. We can only speculate on why feature retrocue benefits were not observed. One difference that could be relevant is that studies obtaining feature retrocue benefits had a lower memory load (2 or 3 items) than used here (6 items) with younger participants, and somewhat longer postcue times (2000–5000 ms), whereas in our study, this delay was of 1500 ms. Future studies are needed to understand under which conditions participants are able to use feature-based attention to prioritize relevant features of objects already stored in visual WM.

In summary, Experiment 2 provided evidence against the hypothesis that older adults have a deficit in directing feature-based attention to representations in visual WM compared to younger adults.

5.3. *Aging Deficits in the Control of Attention?*

A deficit in using visual attention to control visual WM contents has been proposed as one of the main contender explanations for visual WM impairments in aging [45]. This hypothesis predicts that older adults exhausts their WM capacity with irrelevant information, thereby performing more poorly for tests of the relevant information. What is the evidence supporting this hypothesis? Some neurophysiological studies have pointed to an age-related deficit in selective attention as revealed by comparison of brain activations vs. deactivations in face of requirements to attend vs. filter/ignore information [4,46–48] or time for onset of suppression markers in ERP components [16]. Behaviorally, however, the evidence is more challenging for the inhibition hypothesis. Selective attention is beneficial for encoding and maintaining relevant content in visual WM as revealed by precue and retrocue benefits, respectively (for a review see [49]). Studies investigating age-related deficits in the use of selective attention through attentional cues provide challenging results to this inhibition-deficit hypothesis. Some studies showed no age-related deficits in precue performance [50,51], which is akin to the filtering ability investigated in the present study (Exp. 1) and in the study of McNab et al. [9]. In conjunction, these studies and ours show that older adults do not have difficulties in focusing their attention in perceptually available relevant visual information, thereby gating them access to visual WM, and hindering access to distractor information. These results resonate with findings questioning age-related deficits in executive control more generally, which use visual perceptual tasks coupled with distraction manipulations, such as Stroop, Flanker, or Simon tasks [52–54]. This literature has revealed that patterns of age-related decline can be fully explained by reductions in processing speed. Results from retrocue paradigms, conversely, are akin to the use of attention to ignore an irrelevant content maintained in visual WM [55]. Although some initial studies pointed to age-deficits in using retrocues [56,57], subsequent studies have consistently obtained retrocue benefits of similar magnitude between younger and older adults [50,58–61]. Together with the present findings, more and more behavioral evidence indicates that the source of the robust age-related decline observed in visual WM is unlikely to be explained by deficits in inhibiting irrelevant information.

5.4. *Limitations*

In the present study, we only investigated the contribution of visual attention to visual WM performance. Our results therefore may not generalize to verbal working memory tasks, which may involve different brain areas and different mechanisms to gate and maintain information in mind [62].

Another limitation is that we only investigated healthy aging. Filtering and ignoring distractors in visual WM may be impaired in nonhealthy aging (e.g., dementia) in addition to the visual WM deficits that have been found in several neurodegenerative diseases [63,64]. Our results may serve as a benchmark, however, when assessing the impact of nonhealthy changes that can occur in aging.

Lastly, our sample size (N = 30 per group) was modest, and our age range limited to elderly between 65 and 79 years old (M = 72.6), hence not covering old-old age. Future studies may include larger sample sizes and broad age ranges to more fully understand the impact of aging on the relation between attention and visual WM.

6. Conclusions

The current study investigated age-related differences in filtering and ignoring irrelevant objects and irrelevant features of an object. Although inefficient control of attention is usually assumed to underlie age differences in WM capacity, across two experiments, we found no support for the assumption that the older adults are less able than younger adults to filter and ignore distraction in visual WM when selecting different objects or features within an object. Hence, their lower visual WM capacity cannot be explained by the inefficient use of attention to control the contents of visual WM.

Author Contributions: Conceptualization, M.R.M. and A.S.S.; methodology, software, and visualization, A.S.S.; formal analysis, M.R.M. and A.S.S.; investigation, M.R.M.; resources, A.S.S.; data curation, M.R.M. and A.S.S.; writing—original draft preparation, M.R.M.; writing—review and editing, supervision, project administration, and funding acquisition, A.S.S. All authors have read and agreed to the published version of the manuscript.

Funding: This research was funded by a grant from the Velux foundation (project N° 1053) to A.S.S and from CAPES Foundation, Ministry of Education of Brazil to M.R.M.

Acknowledgments: We acknowledge Elena Throm for her assistance with data collection.

Conflicts of Interest: The authors declare no conflict of interest.

References

1. Park, D.C.; Lautenschlager, G.; Hedden, T.; Davidson, N.S.; Smith, A.D.; Smith, P.K. Models of visuospatial and verbal memory across the adult life span. *Psychol. Aging* **2002**, *17*, 299–320. [CrossRef] [PubMed]
2. Salthouse, T.A. When does age-related cognitive decline begin? *Neurobiol. Aging* **2009**, *30*, 507–514. [CrossRef] [PubMed]
3. Bopp, K.L.; Verhaeghen, P. Aging and verbal memory span: A meta-analysis. *J. Gerontol. B Psychol. Sci. Soc. Sci.* **2005**, *60*, P223–P233. [CrossRef] [PubMed]
4. Gazzaley, A.; Clapp, W.; Kelley, J.; McEvoy, K.; Knight, R.T.; D'Esposito, M. Age-related top-down suppression deficit in the early stages of cortical visual memory processing. *Proc. Natl. Acad. Sci. USA* **2008**, *105*, 13122–13126. [CrossRef] [PubMed]
5. Bonnefond, M.; Jensen, O. Alpha Oscillations Serve to Protect Working Memory Maintenance against Anticipated Distracters. *Curr. Biol.* **2012**, *22*, 1969–1974. [CrossRef] [PubMed]
6. Dolcos, F.; Miller, B.; Kragel, P.; Jha, A.; Mccarthy, G. Regional brain differences in the effect of distraction during the delay interval of a working memory task. *Brain Res.* **2007**, *1152*, 171–181. [CrossRef]
7. McNab, F.; Dolan, R.J. Dissociating distractor-filtering at encoding and during maintenance. *J. Exp. Psychol. Hum. Percept. Perform.* **2014**, *40*, 960–967. [CrossRef]
8. Robison, M.K.; Miller, A.L.; Unsworth, N. Individual differences in working memory capacity and filtering. *J. Exp. Psychol. Hum. Percept. Perform.* **2018**, *44*, 1038–1053. [CrossRef]
9. McNab, F.; Zeidman, P.; Rutledge, R.B.; Smittenaar, P.; Brown, H.R.; Adams, R.A.; Dolan, R.J. Age-related changes in working memory and the ability to ignore distraction. *Proc. Natl. Acad. Sci. USA* **2015**, *112*, 6515. [CrossRef]
10. Park, Y.E.; Sy, J.L.; Hong, S.W.; Tong, F. Reprioritization of Features of Multidimensional Objects Stored in Visual Working Memory. *Psychol. Sci.* **2017**, *28*, 1773–1785. [CrossRef]
11. Marshall, L.; Bays, P.M. Obligatory encoding of task-irrelevant features depletes working memory resources. *J. Vis.* **2013**, *13*, 21. [CrossRef]
12. Oberauer, K. Working Memory and Attention—A Conceptual Analysis and Review. *J. Cogn.* **2019**, *2*. [CrossRef] [PubMed]
13. Vogel, E.K.; McCollough, A.W.; Machizawa, M.G. Neural measures reveal individual differences in controlling access to working memory. *Nature* **2005**, *438*, 500–503. [CrossRef] [PubMed]
14. Salthouse, T.A. Aging and measures of processing speed. *Biol. Psychol.* **2000**, *54*, 35–54. [CrossRef]
15. Jost, K.; Mayr, U. Switching between filter settings reduces the efficient utilization of visual working memory. *Cogn. Affect. Behav. Neurosci.* **2016**, *16*, 207–218. [CrossRef] [PubMed]
16. Jost, K.; Bryck, R.L.; Vogel, E.K.; Mayr, U. Are Old Adults Just Like Low Working Memory Young Adults? Filtering Efficiency and Age Differences in Visual Working Memory. *Cereb. Cortex* **2010**, *21*, 1147–1154. [CrossRef]
17. Oberauer, K.; Eichenberger, S. Visual working memory declines when more features must be remembered for each object. *Mem. Cognit.* **2013**, *41*, 1212–1227. [CrossRef] [PubMed]

18. Palmer, J.; Boston, B.; Moore, C.M. Limited capacity for memory tasks with multiple features within a single object. *Atten. Percept. Psychophys.* **2015**, *77*, 1488–1499. [CrossRef]
19. Fougnie, D.; Asplund, C.L.; Marois, R. What are the units of storage in visual working memory? *J. Vis.* **2010**, *10*, 27. [CrossRef]
20. Brockmole, J.R.; Logie, R.H. Age-related change in visual working memory: A study of 55,753 participants aged 8–75. *Front. Psychol.* **2013**, *4*. [CrossRef]
21. Brockmole, J.R.; Parra, M.A.; Sala, S.D.; Logie, R.H. Do binding deficits account for age-related decline in visual working memory? *Psychon. Bull. Rev.* **2008**, *15*, 543–547. [CrossRef] [PubMed]
22. Luck, S.J.; Vogel, E.K. The capacity of visual working memory for features and conjunctions. *Nature* **1997**, *390*, 279–281. [CrossRef] [PubMed]
23. Hardman, K.O.; Cowan, N. Remembering complex objects in visual working memory: Do capacity limits restrict objects or features? *J. Exp. Psychol. Learn. Mem. Cogn.* **2015**, *41*, 325–347. [CrossRef] [PubMed]
24. Shen, M.; Tang, N.; Wu, F.; Shui, R.; Gao, Z. Robust object-based encoding in visual working memory. *J. Vis.* **2013**, *13*, 1. [PubMed]
25. Yin, J.; Zhou, J.; Xu, H.; Liang, J.; Gao, Z.; Shen, M. Does high memory load kick task-irrelevant information out of visual working memory? *Psychon. Bull. Rev.* **2012**, *19*, 218–224. [CrossRef]
26. Shin, H.; Ma, W.J. Crowdsourced single-trial probes of visual working memory for irrelevant features. *J. Vis.* **2016**, *16*, 10. [CrossRef]
27. Yu, Q.; Shim, W.M. Occipital, parietal, and frontal cortices selectively maintain task-relevant features of multi-feature objects in visual working memory. *NeuroImage* **2017**, *157*, 97–107. [CrossRef]
28. Serences, J.T.; Ester, E.F.; Vogel, E.K.; Awh, E. Stimulus-specific delay activity in human primary visual cortex. *Psychol. Sci.* **2009**, *20*, 207–214. [CrossRef]
29. Niklaus, M.; Nobre, A.C.; van Ede, F. Feature-based attentional weighting and spreading in visual working memory. *Sci. Rep.* **2017**, *7*. [CrossRef]
30. Ye, C.; Hu, Z.; Ristaniemi, T.; Gendron, M.; Liu, Q. Retro-dimension-cue benefit in visual working memory. *Sci. Rep.* **2016**, *6*, 35573. [CrossRef]
31. Brainard, D.H. The psychophysics toolbox. *Spat. Vis.* **1997**, *10*, 433–436. [CrossRef] [PubMed]
32. Pelli, D.G. The VideoToolbox software for visual psychophysics: Transforming numbers into movies. *Spat. Vis.* **1997**, *10*, 437–442. [CrossRef] [PubMed]
33. Zhang, W.; Luck, S.J. Discrete fixed-resolution representations in visual working memory. *Nature* **2008**, *453*, 233–235. [CrossRef] [PubMed]
34. Kruschke, J.K. Bayesian Assessment of Null Values Via Parameter Estimation and Model Comparison. *Perspect. Psychol. Sci.* **2011**, *6*, 299–312. [CrossRef]
35. Rouder, J.N.; Morey, R.D.; Speckman, P.L.; Province, J.M. Default Bayes factors for ANOVA designs. *J. Math. Psychol.* **2012**, *56*, 356–374. [CrossRef]
36. R Core Team. R: A Language and Environment for Statistical Computing. Available online: https://www.gbif.org/tool/81287/r-a-language-and-environmentfor-statistical-computing (accessed on 13 August 2020).
37. Morey, R.D.; Rouder, J.N. BayesFactor: Computation of Bayes Factors for Common Designs. Available online: https://richarddmorey.github.io/BayesFactor/ (accessed on 13 August 2020).
38. Kruschke, J.K.; Meredith, M. Bayesian Estimation Supersedes the *t*-Test. Available online: https://cran.r-project.org/web/packages/BEST/BEST.pdf (accessed on 13 August 2020).
39. Salthouse, T.A. Selective review of cognitive aging. *J. Int. Neuropsychol. Soc. JINS* **2010**, *16*, 754–760. [CrossRef]
40. Hdstes, T.; Poon, L.W.; Cerella, J.; Frzard, J.L. Age-related differences in the time course of encoding. *Exp. Aging Res.* **1982**, *8*, 175–178. [CrossRef]
41. Zanto, T.P.; Toy, B.; Gazzaley, A. Delays in neural processing during working memory encoding in normal aging. *Neuropsychologia* **2010**, *48*, 13–25. [CrossRef]
42. Souza, A.S.; Czoschke, S.; Lange, E.B. Gaze-based and attention-based rehearsal in spatial working memory. *J. Exp. Psychol. Learn. Mem. Cogn.* **2020**, *46*, 980–1003. [CrossRef]
43. Maunsell, J.H.R.; Treue, S. Feature-based attention in visual cortex. *Trends Neurosci.* **2006**, *29*, 317–322. [CrossRef]

44. Woodman, G.F.; Vogel, E.K. Selective storage and maintenance of an object's features in visual working memory. *Psychon. Bull. Rev.* **2008**, *15*, 223–229. [CrossRef] [PubMed]
45. Hasher, L.; Zacks, R.T. Working Memory, Comprehension, and Aging: A Review and a New View. In *Psychology of Learning and Motivation*; Bower, G.H., Ed.; Academic Press: Cambridge, MA, USA, 1988; pp. 193–225.
46. Gazzaley, A.; Cooney, J.W.; Rissman, J.; D'Esposito, M. Top-down suppression deficit underlies working memory impairment in normal aging. *Nat. Neurosci.* **2005**, *8*, 1298–1300. [CrossRef] [PubMed]
47. Chao, L.; Knight, R.; Chao, L.L.; Knight, R.T. Prefrontal deficits in attention and inhibitory control with aging. *Cereb. Cortex N. Y. N. 1991* **1997**, *7*, 63–69. [CrossRef] [PubMed]
48. West, R.; Alain, C. Age-related decline in inhibitory control contributes to the increased Stroop effect in older adults. *Psychophysiology* **2000**, *37*, 179–189. [CrossRef] [PubMed]
49. Souza, A.S.; Oberauer, K. In search of the focus of attention in working memory: 13 years of the retro-cue effect. *Atten. Percept. Psychophys.* **2016**, *78*, 1839–1860. [CrossRef] [PubMed]
50. Souza, A.S. No age deficits in the ability to use attention to improve visual working memory. *Psychol. Aging* **2016**, *31*, 456–470. [CrossRef]
51. Curtis, A.; Turner, G.; Park, N.; Murtha, S.J.E. Improving visual spatial working memory in younger and older adults: Effects of cross-modal cues. *Aging Neuropsychol. Cogn.* **2017**, *26*, 1–20. [CrossRef]
52. Verhaeghen, P.; Cerella, J. Aging, executive control, and attention: A review of meta-analyses. *Neurosci. Biobehav. Rev.* **2002**, *26*, 849–857. [CrossRef]
53. Rey-Mermet, A.; Gade, M.; Oberauer, K. Should we stop thinking about inhibition? Searching for individual and age differences in inhibition ability. *J. Exp. Psychol. Learn. Mem. Cogn.* **2018**, *44*, 501–526. [CrossRef]
54. Rey-Mermet, A.; Gade, M. Inhibition in aging: What is preserved? What declines? A meta-analysis. *Psychon. Bull. Rev.* **2018**, *25*, 1695–1716. [CrossRef]
55. Lepsien, J.; Nobre, A.C. Cognitive control of attention in the human brain: Insights from orienting attention to mental representations. *Control. Atten. Actions* **2006**, *1105*, 20–31. [CrossRef] [PubMed]
56. Newsome, R.N.; Duarte, A.; Pun, C.; Smith, V.M.; Ferber, S.; Barense, M.D. A retroactive spatial cue improved VSTM capacity in mild cognitive impairment and medial temporal lobe amnesia but not in healthy older adults. *Neuropsychologia* **2015**, *77*, 148–157. [CrossRef] [PubMed]
57. Duarte, A.; Hearons, P.; Jiang, Y.; Delvin, M.C.; Newsome, R.N.; Verhaeghen, P. Retrospective attention enhances visual working memory in the young but not the old: An ERP study. *Psychophysiology* **2013**, *50*, 465–476. [CrossRef] [PubMed]
58. Gilchrist, A.L.; Duarte, A.; Verhaeghen, P. Retrospective cues based on object features improve visual working memory performance in older adults. *Aging Neuropsychol. Cogn.* **2016**, *23*, 184–195. [CrossRef]
59. Mok, R.M.; Myers, N.E.; Wallis, G.; Nobre, A.C. Behavioral and Neural Markers of Flexible Attention over Working Memory in Aging. *Cereb. Cortex* **2016**, *26*, 1831–1842. [CrossRef] [PubMed]
60. Loaiza, V.M.; Souza, A.S. Is refreshing in working memory impaired in older age? Evidence from the retro-cue paradigm. *Ann. N. Y. Acad. Sci.* **2018**, *1424*, 175–189. [CrossRef]
61. Loaiza, V.M.; Souza, A.S. An age-related deficit in preserving the benefits of attention in working memory. *Psychol. Aging* **2019**, *34*, 282–293. [CrossRef]
62. Bopp, K.L.; Verhaeghen, P. Age-Related Differences in Control Processes in Verbal and Visuospatial Working Memory: Storage, Transformation, Supervision, and Coordination. *J. Gerontol. Ser. B* **2007**, *62*, P239–P246. [CrossRef]
63. Zokaei, N.; Husain, M. Working Memory in Alzheimer's Disease and Parkinson's Disease. In *Processes of Visuospatial Attention and Working Memory*; Hodgson, T., Ed.; Springer International Publishing: Cham, Switzerland, 2019; pp. 325–344.
64. Cecchini, M.A.; Yassuda, M.S.; Bahia, V.S.; de Souza, L.C.; Guimaraes, H.C.; Caramelli, P.; Carthery-Goulart, M.T.; Patrocinio, F.; Foss, M.P.; Tumas, V.; et al. Recalling feature bindings differentiates Alzheimer's disease from frontotemporal dementia. *J. Neurol.* **2017**, *264*, 2162–2169. [CrossRef]

Article

Reduced Attentional Control in Older Adults Leads to Deficits in Flexible Prioritization of Visual Working Memory

Sarah E. Henderson †, Holly A. Lockhart †, Emily E. Davis, Stephen M. Emrich and Karen L. Campbell *

Department of Psychology, Brock University, St Catharines, ON L2S 3A1, Canada; sh14jm@brocku.ca (S.E.H.); hl10ze@brocku.ca (H.A.L.); ed11zq@brocku.ca (E.E.D.); semrich@brocku.ca (S.M.E.)

* Correspondence: karen.campbell@brocku.ca

† S.E.H. and H.A.L. contributed equally to this work.

Received: 30 June 2020; Accepted: 8 August 2020; Published: 11 August 2020

Abstract: Visual working memory (VWM) resources have been shown to be flexibly distributed according to item priority. This flexible allocation of resources may depend on attentional control, an executive function known to decline with age. In this study, we sought to determine how age differences in attentional control affect VWM performance when attention is flexibly allocated amongst targets of varying priority. Participants performed a delayed-recall task wherein item priority was varied. Error was modelled using a three-component mixture model to probe different aspects of performance (precision, guess-rate, and non-target errors). The flexible resource model offered a good fit to the data from both age groups, but older adults showed consistently lower precision and higher guess rates. Importantly, when demands on flexible resource allocation were highest, older adults showed more non-target errors, often swapping in the item that had a higher priority at encoding. Taken together, these results suggest that the ability to flexibly allocate attention in VWM is largely maintained with age, but older adults are less precise overall and sometimes swap in salient, but no longer relevant, items possibly due to their lessened ability to inhibit previously attended information.

Keywords: aging; working memory; visual working memory; attentional control

---

1. Introduction

The ability to maintain visual representations in working memory is critical in our dynamic environment, enabling the integration of events and information over time to facilitate decision-making and other higher-order executive processes. Unfortunately, working memory performance tends to decline with age [1–3], which may be attributed to age-related deficits in attentional control [4,5]. Recent models of visual working memory (VWM) suggest that attentional resources are flexibly allocated across stimuli according to their relative priority [6] rather than assigned to a set number of slots in working memory. Further, the ability to flexibly prioritize certain items in memory appears to rely on attentional control [7]. Thus, we might expect older adults to be less flexible in their allocation of attention, but this has yet to be explored in visual working memory.

1.1. VWM Resources Are Flexibly Allocated

The amount of information that can be held in working memory is limited [8]; as storage demands increase, either the fidelity of the representations in memory, or the number of maintained representations, decreases [9,10]. Two major theories are often used to explain the limits of working memory capacity: the discrete-capacity [9] and continuous-resource models [11]. The discrete-capacity model argues that there exist a fixed number of items that can be encoded with high fidelity in storage "slots", with any items beyond that number failing to be encoded [12]. In opposition to this model is the continuous-resource model, which posits that working memory is not constrained by the number of representations that can be maintained, but rather by a limited pool of resources that can be distributed across items, such that lesser resource allocation per item results in lower-fidelity representations [11,13]. Thus, the continuous-resource model predicts that there is greater flexibility in the allocation of resources in working memory. However, most of this work has examined changes in performance by manipulating memory load, assuming equal division of resources between items, and largely ignoring the possibility that some items may be prioritized over others in accordance with one's goals.

The flexible nature of working memory has been illustrated through work suggesting that the resources allocated to each item in working memory may not always be constant. Factors such as random fluctuations in attention at encoding [14], attentional cuing [12], rewards [15], and voluntary control (at least when load is low) [16] have all been shown to allow for between-item variability in quality of representation. Recent models of visual working memory have applied the continuous-resource model with the aim of assessing the flexibility of VWM resources more directly. Emrich and colleagues [6] used cues to manipulate the relative priority of different memoranda to examine whether spatial attention divided between items could account for error in VWM. Performance was found to be better predicted by the prioritization assigned to each item than by memory load alone. This relationship between attentional priority and memory performance followed a power law, suggesting that memory resources can be flexibly and continuously allocated to visual stimuli depending on attentional prioritization and, importantly, providing evidence that working memory performance can be improved through appropriate allocation of VWM resources [6,17].

1.2. Flexible Allocation Relies on Attentional Control

A flexible allocation of attention account of VWM is consistent with numerous models of VWM that identify attention as a critical factor in determining the information that is successfully encoded in working memory stores [18–21]. There are two distinct mechanisms through which attentional control may influence working memory encoding: through filtering/inhibiting distractors [22] and through biasing attention towards target items independent of filtering [6,15]. Filtering efficiency can be quantified using contralateral delay activity (CDA)—an event-related electrophysiological measure of sustained activity observed during the delay period of a memory task [9,23]. Increases in the number or complexity of items stored in VWM are associated with increased CDA amplitude, which saturates and appears to plateau when memory load reaches a few items [24]. However, low-capacity individuals show greater CDA amplitude than high-capacity individuals when distractors are present [25], reflecting storage of irrelevant items (i.e., poor filtering efficiency), and this relates to poorer target recall. Further, the association between poor attentional control and low VWM capacity has been linked to difficulty in overriding attentional capture, where low-capacity individuals have greater difficulty controlling spatial attention in the presence of distractors [26]. Thus, filtering efficiency appears to be a critical determinant of VWM performance [25] (although see [27]).

Similarly, the flexible prioritization of resources in working memory through attentional biasing can be tracked by a separate earlier electrophysiological component, the N2pc. Greater negativity of the N2pc is thought to reflect enhancement of task-relevant information prior to memory maintenance [28,29]. A larger N2pc has been observed for memory items prioritized by probabilistic cues with greater negative amplitude associated with higher-probability cues suggesting greater attentional enhancement of more probable items [7]. Thus, the flexible allocation of attention relies on both the suppression of distractors and the enhancement of task-relevant stimuli.

1.3. Aging, VWM, and Attentional Control

Consistent with attentional accounts of VWM performance, it has been established that VWM performance declines with age [3,30–32], and this is thought to be caused by concomitant declines in attentional control [4]. While older adults can sometimes use valid spatial cues to improve VWM performance in a manner similar to younger adults, they still display worse working memory overall [33]. This is particularly the case when distracting information is present. Past work suggests that older adults show impaired suppression of distracting information and are, thus, more likely to let that information into VWM, which in turn affects target recall [31,34,35]. Indeed, age-related increases in susceptibility to distraction have been linked to prolonged processing of distracting information [36]. This is the case even when distracting information is completely irrelevant to the task, as when auditory distractors were shown to have deleterious effects on visual working memory in older but not younger adults [37]. This work suggests that deficits in filtering efficiency or inhibition of irrelevant information can at least partially explain age differences in VWM [38–40] (although see [41]).

Recent work using mixture models to characterize VWM performance has provided further insight into the nature of age-related declines in VWM. These mixture models can identify the specific types of errors that participants make in a memory recall task, as well as define the precision or fidelity of visual representations in working memory. Using continuous response to measure memory for colour and orientation, Peich and colleagues [3] reported an age-related decline in VWM precision. They also showed an age-related increase in non-target errors or a greater tendency to report unprobed items from the original memory array, suggesting an age-related deficit in feature binding [42–44]. Similar findings were observed in a large, population-based sample who showed age-related declines in the fidelity and quantity of remembered items (coloured circles) [45]. Taken together, this work suggests that working memory representations become less precise with age, at least when attention is allocated equally across memoranda. It remains to be seen, however, whether older adults can flexibly allocate their attention when some items are prioritized over others in VWM.

1.4. Current Study

In this study, we aimed to examine whether older adults can flexibly allocate attentional resources in VWM in response to probabilistic cues. Using a delayed-recall paradigm, participants viewed coloured squares with probabilistic spatial cues that indicated the likelihood that each item would be probed (and thus, its priority), and then reported one of the squares using a continuous response (see Figure 1). The cues varied in number and validity across conditions such that the attentional resources to be allocated to each item would vary between 16.7% and 100% while maintaining a constant set size (except in a baseline single-item condition). We hypothesized that: (1) a flexible-allocation account would explain response error in younger adults [6] and that this pattern would largely hold for older adults; (2) older adults would show greater error overall reflecting a general decline in visual working memory (fidelity and capacity) [3]; and (3) older adults would have particular difficulty flexibly allocating attention when demands on attentional control are high (i.e., in the highest flexible attention prioritization condition) due to reduced attentional control [4,34,35].

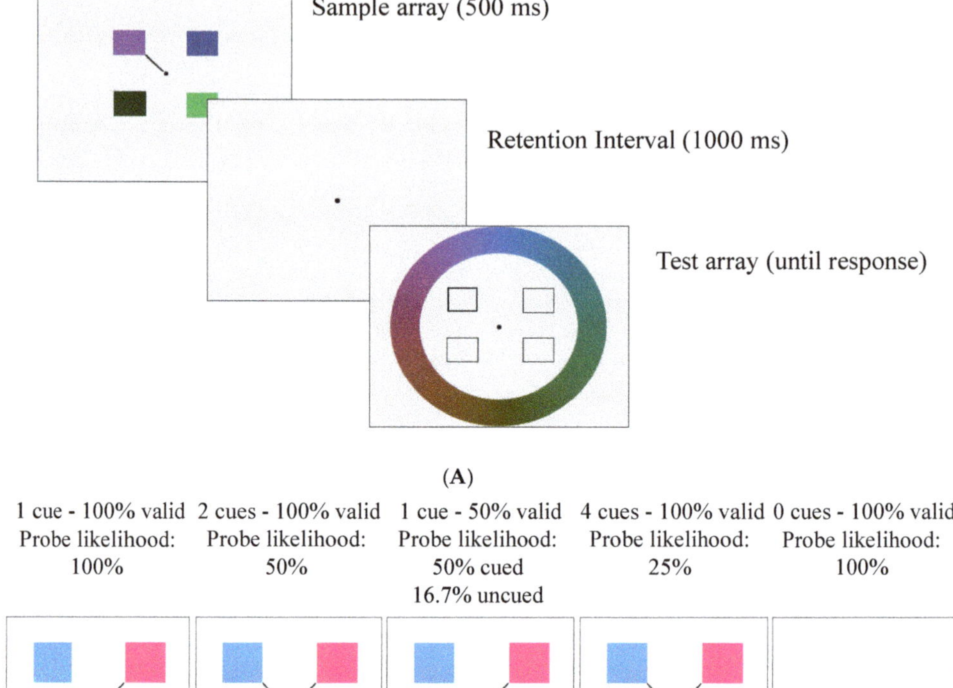

Figure 1. (**A**) Example trial from the delayed-recall task. Participants were shown a sample array for 500 ms with either one (in the 0 cue condition) or 4 squares (with 1, 2, or 4 line cues), which were blocked according to cue number and validity (conditions described in Table 1). Following a 1000 ms retention interval, a dark outline surrounded the square required to be recalled. Participants were required to select a colour on the colour wheel that matched the probed location in the sample array. (**B**) Sample arrays of all the trial conditions.

Table 1. Condition parameters and trial information.

Condition	Number of Items	Number of Cues	Memory Load	Cue Validity (%)	Probability of Cued Item Probed (%)	Probability of Uncued Item Probed (%)	Number of Trials Day 1	Number of Trials Day 2
0 cues–100%	1	0	1	–	–	–	33	67
1 cue–100%	4	1	1	100	100	0	33	67
1 cue–50%	4	1	4	50	50	16.67	33	67
2 cues–100%	4	2	2	100	50	0	33	67
4 cues–100%	4	4	4	100	25	0	66	134

2. Material and Methods

2.1. Participants

Participants included 20 older and 20 younger adults. One older adult was replaced because they scored less than 23 on the Montreal Cognitive Assessment (MoCA) [46,47] and two younger adults were replaced for not completing the second session. Older adults were recruited from a community participant pool and paid $10/h. Younger adults were recruited through a psychology student pool and received course credit for their time. Sample sizes were chosen in line with previous work from which the present study was adapted [6].

Older adults scored higher on the Shipley Vocabulary Test than younger adults, $t(38) = 3.62$, $p < 0.001$, in line with previous work [48]. There was no age difference in years of education (see Table 2 for all demographics). No participants reported being colour blind, and all participants met the criteria for normal colour vision after completing the Ishihara Test of Colour Vision.

Table 2. Demographics.

Group	Age (Years)			Education (Years)			Vocabulary			MoCA		
	M	Range	SD	M	Range	SD	M	Range	SD	M	Range	SD
Younger	21.10	18–26	2.34	15.65	12–20	1.87	0.74	0.60–0.98	0.12	-	-	-
Older	73.32	65–85	4.92	15.25	9–40	6.38	0.87	0.45–1	0.12	25.75	23–29	1.71

MoCA: Montreal Cognitive Assessment. Vocabulary is reported as proportion correct. Age for one older adult participant was missing.

2.2. Procedure and Stimuli

The study was approved by the Research Ethics Board at Brock University. The experiment was split into two, 1 h sessions (1.5 h for older adults) to reduce fatigue. All sessions took place in the afternoon with a maximum of seven days between sessions. All computer tasks were administered with PsychoPy (v.3.0.4) [49] on a 20 inch Dell CRT monitor (1024 × 768 px) with participants approximately 60 cm away from the monitor. In Session 1, after providing informed consent, participants were tested for colour blindness, and then completed three separate computer tasks: change detection (10 min), psychological distance triad (10 min) [50], and delayed recall (1/3 trials). In Session 2, participants completed the rest of the delayed-recall task (2/3 trials), followed by the MoCA, Shipley Vocabulary Test, and a demographics questionnaire (i.e., age, sex, education). Data from the change detection and psychological distance triad tasks are not reported here.

2.3. Delayed-Recall Task

This task was adapted from Emrich and colleagues [6]. Participants viewed four squares (1.8° × 1.8°) around a fixation dot (0.3° diameter) for 500 ms on a grey background. The squares were evenly spaced in four quadrants of the screen with 1, 2, or 4 line cues (see Figure 1). The line cues varied in their validity, and participants were explicitly told the validity of the cues at the start of each block. A total of four cued conditions were presented in separate blocks: 4 cues–100% valid (meaning one of those 4 squares would be tested); 2 cues–100% valid (meaning one of those 2 squares would be tested); 1 cue–50% valid (meaning there is a 50% chance that the cued square will be tested, and a 16.67% chance that each of the other squares will be tested); 1 cue–100% valid. There was also a no-distractor condition (0 cues–100% valid) in which participants viewed a single square that appeared randomly in one of the four quadrants with no cue (see Table 1). In all conditions except the 1 cue–50% valid condition, the uncued items would never be tested and thus could be ignored. For clarity, the probability of a cued item being probed has been termed "probe likelihood".

Following a 1000 ms retention period, the test array appeared. The test array contained four square outlines in the same four locations as the sample array. One of the square outlines had a thicker border than the others indicating that participants had to recall the colour of the square that was in that location in the sample array. Participants selected the colours via a continuous-response colour wheel (14° radius) that appeared around the square outlines and contained all possible colours used in the experiment (see Figure 1). As participants moved a black probe around the colour wheel using the "c" (clockwise) and "m" (counterclockwise) keys, the tested square infilled with the colour of the wheel. When participants felt that the colour in the square was as close as possible to the correct colour, they pressed "space" to lock in their answer. Following participants' decision, there was a 500 ms intertrial interval, then the next trial began. The no-distractor condition was presented in the same way, except that the test array had only one square outline in the same location as the sample array, which only included one square. Participants performed a total of 198 trials during Session 1 and 402 trials in Session 2. During each session, participants saw all five conditions in blocks (a breakdown of the number of trials per condition per day is provided in Table 1).

Sample colours were selected randomly from one of 360 unique colours obtained from a circular wheel on the CIE L*a*b* colour space with coordinates of a = −6 and b = 14 with a radius of 49, calibrated to the monitor, with a minimum distance of 30 degrees between sample colours.

2.4. Analysis

Responses in the delayed-recall task from Session 1 and Session 2 were merged using R. Due to a programming error, the 1 cue–50% valid condition only contained 100 trials, meaning that the uncued locations were only probed on 50 trials, providing fewer opportunities for participants to respond correctly to these trials than in the other conditions. To correct for this, these data were bootstrapped up to 200 trials (100 cued and 100 uncued) in R using the boot package [51,52]. Additionally, the 4 cue–100% trials accidentally included 100 extra trials, so only the first 100 were analysed in order to equate this condition to the other conditions. All reported analyses are based on the corrected data; however, it should be noted that the same pattern of results is observed when analyses are limited to the original (uncorrected) data (see Supplementary Figure S1 and Supplementary Table for uncorrected results).

2.4.1. Response Error

Power law vs. Linear fit. Response error was calculated as the circular distance between the target value and the reported value on each trial, then the standard deviation of response error was calculated for each condition for each participant. The standard deviation of response error was chosen because it captures the precision of responses and is the statistic calculated by the mixture model [13]. A linear mixed-effects model was used to assess whether the relationship between probe likelihood and response error is better explained by a power law [6] relative to a simpler linear fit. This analysis was completed in R using the packages *lme4* [53], *lmerTest* [54], *car* [55], and *sjPlot* [56]. For the linear fit, response error was predicted by a fixed effect of probe likelihood with a random intercept for each participant. This is analogous to a repeated-measures ANOVA, which allows each participant to vary from each other, but the effect of the predictor (probe likelihood) is the same for each participant. Because a power-law function is linear when both the predictor (x) and outcome (y) are log transformed, the log of SD response error was predicted by a fixed effect of the log of probe likelihood with a random intercept for each participant. Degrees of freedom were estimated using the Kenward–Roger approximation, which is known to reduce bias in small-sample parameter estimation in linear models [57]. Model fit is reflected by AIC values (with lower values indicating better fit) and these were compared between models. After finding the best fitting model for each group, the marginal R-squared values were compared between groups using Fisher's Z test for independent sample comparison.

Key comparisons. The hypothesis that attentional resources are flexibly allocated across memoranda leads to two specific predictions. First, items of equivalent probe likelihood (i.e., similar priority) should receive similar amounts of attention and, thus, have similar response error. Specifically, cued items from the 1 cue–50% valid condition should receive the same amount of attention as 2 items in the 2 cues–100% valid condition. To test the hypothesis of no difference between these conditions, response error was submitted to a Bayesian repeated-measures ANOVA with condition (1 cue–50% valid vs. 2 cues–100% valid) as a within-subjects factor and age group (young vs. old) as a between-subjects factor. The second key prediction is that items with different probe likelihood, but which are similarly cued (i.e., different cue validity and thus different priority), will not receive similar amounts of attention and should show differences in response error. To test this hypothesis, response error was submitted to a Bayesian repeated-measures ANOVA with cue validity condition (1 cue–100% valid vs. 1 cue–50% valid) as a within-subjects factor and age group (young vs. old) as a between-subjects factor.

For ease of reading, the BF_{01} is reported when the prediction was in favour of the null hypothesis and BF_{10} when the prediction was against the null hypothesis. These analyses were run in JASP [58]. For reference, a $BF_{01} > 3$ (equivalent of $BF_{10} < 0.33$) suggests moderate evidence for the null hypothesis and $BF_{10} > 3$ suggests moderate evidence for the alternative hypothesis; values greater than 10 are considered strong evidence in the respective directions [59]. Additionally, BF_{incl} refers to the posterior probability that the inclusion of the model term or interaction would produce a model that explains the observed data. Models that include the interaction term always include the lower-order terms, so where inclusion of the interaction term is justified, only the interaction BF_{incl} is reported. Evidence for the exclusion of the condition term ($BF_{incl} < 0.33$) suggests no difference between conditions and in these cases, the best supported model does not include the condition term.

Spatial Cue Utilization. Although not a prediction of the flexible-allocation account of memory and attention resources, we included a no-distractor condition that could verify whether any observed age differences were due to a catastrophic failure to use the spatial cues. The 0-cue and 1-cue–100% conditions both require a single item to be remembered, but the 1 cue–100% condition also includes 3 distracting items that should be ignored. By comparing the 1 cue–100% valid to the no-distractor condition, we can determine whether both groups are able to use spatial cues to effectively filter distractors under low flexible-allocation demands. Response error from these conditions was submitted to a Bayesian repeated-measures ANOVA with condition (1 cue–100% valid vs. no-distractor) as a within-subjects factor and age group (young vs. old) as a between-subjects factor. Performance should be high and equivalent in both conditions if participants are able to use a highly predictive cue to direct attention.

2.4.2. Mixture Model

Response error in the delayed-recall task was split by condition and modelled using the three-component mixture model [13] in MemToolbox for MATLAB R2017a [60]. This mixture model gives output parameters for guess rate (proportion of random responses), non-target rate (proportion of responses centred around one of the non-target items), and response precision (the standard deviation of the circular von Mises distribution, centred around the target value) in degrees. The approximate number of trials used to estimate the precision parameter (correct trials) can be calculated according to the following formula: trials = (total trials) × (1-guess rate + non-target rate). Since precision is based on the circular normal (von Mises) distribution of response error after accounting for a uniform distribution containing guesses and non-target responses, this method means precision is calculated on correct trials only, but because of the continuous-response method, correct/incorrect is not a binary decision and is independent for each participant. If there were under 10 trials used to calculate a participant's precision in any condition, the modelled precision output for that condition was removed from the analysis. This criterion was determined so that the precision parameter would be a reliable estimate while retaining as many older participants as possible. This criterion resulted in the removal of the uncued response data from the 1 cue–50% valid condition in 1 young adult and 5 older adults. After excluding these participants, the average number of correct trials in young adults was 60.1 with a mean precision of 26.6 for this condition, while older adults had an average of 40.9 correct trials with a mean precision of 31.6.

Parameter estimates (guess rate, non-target rate, and precision) were submitted to separate Bayesian repeated-measures ANOVAs with probe likelihood (100%, 50%, 25%, 16.67%) as a within-subjects factor and age group (young vs. old) as a between-subjects factor. Models for each factor and the interactions were compared to a null model. These analyses were run in JASP [58].

3. Results

3.1. Response Error

We first examined raw response error to compare whether older and younger adults could flexibly allocate VWM resources in a similar way. Previous work with younger adults has shown that response error is affected by probe likelihood in a continuous fashion according to a power law [6,7,17]; thus, our first question was whether older adults' response error could similarly be characterized by a power law. To this end, a linear mixed-effect model was run on the log-transformed response error data with log-transformed probe likelihood as a predictor, and this was compared to a simpler linear fit (i.e., untransformed variables). The analysis was first run on the whole sample with group as a predictor and then run on each group independently. The whole-group analysis revealed a significant group difference such that older adults had greater error overall, ($t(42.11) = 5.40$, $p < 0.001$). The power law offered a better fit to the data than the linear fit ($AIC_{power\ law} = -11.11$ vs. $AIC_{lineaar} = 7450.42$) in the sample as a whole and also in the younger ($AIC_{power\ law} = 5.63$ vs. $AIC_{linear} = 3511.97$) and older groups alone ($AIC_{power\ law} = 7.29$ vs. $AIC_{linear} = 3857.24$). To test whether the power law offered a better fit to the data in either group, a Fisher's Z test compared the marginal R-squared values between groups. While the fit was numerically better in young adults, the difference was not significant ($R^2_{Older\ Adults} = 0.388$, $R^2_{Young\ Adults} = 0.429$, $Z = 0.158$, $p = 0.437$). Thus, both younger and older adults appear to use the attentional cues to flexibly allocate VWM resources.

3.1.1. Key Comparisons

Supporting a flexible-allocation account, both groups reported items of equal priority with equal error (i.e., 1 cue–50% valid vs. 2 cues–100% valid; see Supplementary Figure S1). A Bayesian repeated-measures ANOVA only found support for the inclusion of group ($BF_{incl} = 695.69$) and moderate support for the exclusion of condition (1 cue–50% valid vs. 2 cues–100% valid; $BF_{incl} = 0.32$), with no support either way for the interaction term ($BF_{incl} = 0.77$); thus, there was strong evidence that the best model was a group-only model (Model $BF_{10} = 868.85$). Analysis of the group difference showed strong evidence that older adults reported more error overall ($BF_{10} > 1000$). In other words, although older adults were worse overall, they showed a similar ability to flexibly allocate their attention according to the cue validity.

Further, both groups seem to differentiate a single-cued item by cue validity (i.e., 1 cue–100% valid vs. 1 cue–50% valid; see Supplementary Figure S1). A Bayesian repeated-measures ANOVA found strong evidence for the inclusion of the interaction of age group and cue validity (interaction $BF_{incl} = 25.20$); there was strong support for a model that included group, cue validity, and their interaction (Model $BF_{10} > 1000$). Analysis of the group difference showed there was strong evidence that older adults report greater error overall ($BF_{10} > 1000$), and that there was greater error in the 1 cue–50% valid condition than in the 100% valid condition ($BF_{10} > 1000$). To follow-up the interaction, Bayesian independent samples t-tests showed a larger group difference in the 50% valid ($BF_{10} = 319.37$) than the 100% valid ($BF_{10} = 109.74$) single-cue condition. Taken together, these key comparison results suggest that participants were using cue validity to appropriately allocate memory and attentional resources.

3.1.2. Spatial Cue Utilization

Although the previous analyses suggest both groups were able to use spatial cues to allocate attention and memory resources, we further examined older adults' ability to use spatial cues by comparing the 0-cue (no-distractor) condition to the 1 cue–100% valid condition. The no-distractor and 1 cue–100% both require a single item to be remembered, but the 1 cue–100% condition also includes 3 distracting items that should be ignored. Distributions of the response errors for each condition by older and young adults can be seen in Figure 2A. A Bayesian repeated-measures ANOVA found best support for a group-only model (Model $BF_{10} = 438.36$). There was moderate support for no effect of condition, given the evidence for exclusion of condition term (BF_{incl} = 0.32). Additionally, there was no support either way for the inclusion of the interaction between age group and condition (BF_{incl} = 0.42). Follow-up Bayesian paired sample t-tests within each group found moderate evidence of no difference between conditions (Older Adults, BF_{01} = 8.414; Young Adults, BF_{01} = 4.341). This suggests that both groups were able to effectively use the spatial cues to direct attention and ignore distractors when a single item was cued with 100% validity. While analysis of the group difference again showed strong evidence that older adults had higher error overall (BF_{10} > 1000).

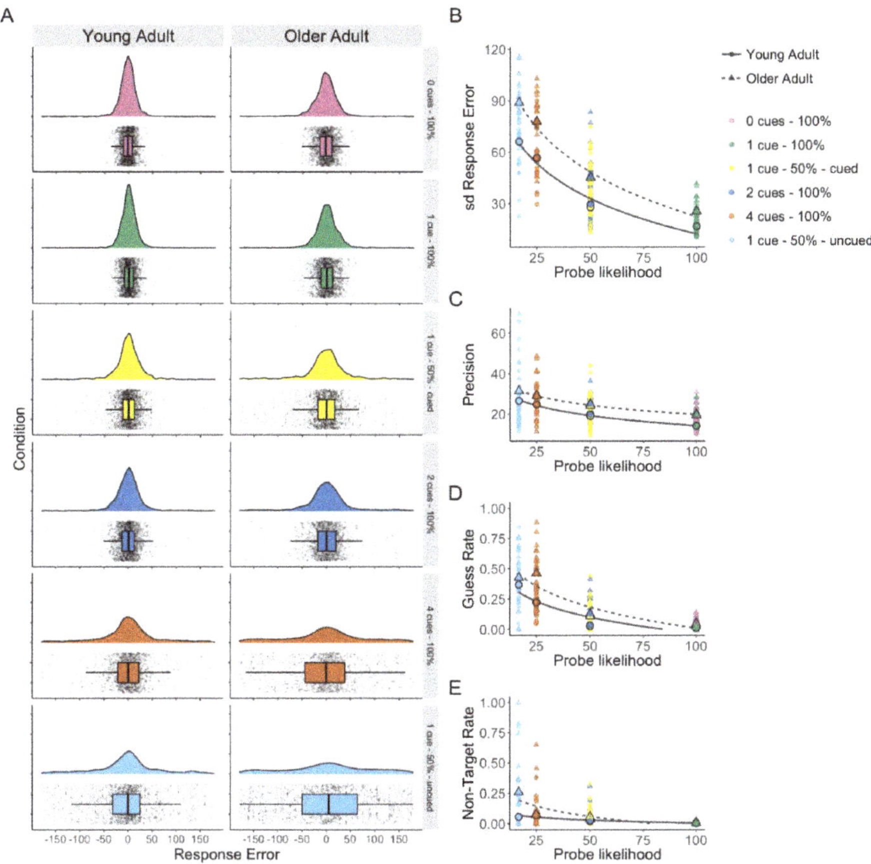

Figure 2. (**A**) Density plots of response error (distance from the target colour in degrees) by condition with box plots and all data points shown for older and young adults (based on raincloud-plots [61]). (**B**) Response error (SD) by probe likelihood. (**C**) Mixture model precision (error in degrees from probed colour) by probe likelihood. (**D**) Mixture model guess rate by probe likelihood. (**E**) Mixture model non-target response rate by probe likelihood.

3.2. Mixture Model

The three-component mixture model enables the assessment of age-related changes in separate aspects of working memory performance using parameter estimates of guess rate, non-target error rate, and precision. Flexible allocation of attention predicts that precision will decrease as attention allocated to the probed stimuli decreases [6]. Further, older adults were expected to have lower precision overall [3]. A Bayesian repeated-measures ANOVA found the best support for a model including probe likelihood and group (both $BF_{incl} > 6$) but excluding the interaction term ($BF_{incl} = 0.11$), Model $BF_{10} > 1000$. Analysis of the group difference showed strong evidence that older adults were less precise ($BF_{10} = 187.89$; see Figure 2C), and this was consistent across the probe likelihood conditions.

Guess rate was expected to increase as probe likelihood decreased, reflecting poorer memory for less attended items across both age groups. Additionally, older adults were expected to show higher guess rates overall, reflecting lower visual working memory abilities. A Bayesian repeated-measures ANOVA found the strongest support for the model including a probe likelihood × group interaction ($BF_{incl} = 54.09$), Model $BF_{10} > 1000$ (see Figure 2D). Analysis of the group difference showed strong evidence that older adults generally had higher guess rates ($BF_{10} = 19.44$). To investigate the interaction, Bayesian independent sample t-tests were performed. There was no evidence of a group difference at either the highest probe likelihood (1 cue–100% valid) or the lowest probe likelihood (1 cue–50% uncued); however, there was a pattern of increasingly larger differences between groups as probe likelihood decreased from the 0 cue–100% valid condition ($BF_{10} = 3.61$) to 1 cue–50% valid cued ($BF_{10} = 5.92$), 2 cues–100% valid ($BF_{10} = 15.64$), and as probe likelihood further decreased there is strongest evidence for a difference in 4 cues–100% valid ($BF_{10} = 139.98$).

Finally, non-target error rate (or "swap rate") was expected to increase in older but not younger adults when flexible prioritization demands increased, but memory load remained constant.

A Bayesian repeated-measures ANOVA found the strongest support for the model including a probe likelihood × group interaction (interaction $BF_{incl} = 126.884$), Model $BF_{10} > 1000$ (see Figure 2E). To investigate the interaction, Bayesian independent sample t-tests were performed. Only the 1 cue–50% uncued condition showed moderate evidence for a difference between age groups ($BF_{10} = 4.03$), all other conditions showed inconclusive evidence (BF_{10}s < 0.48) except the 1 cue–100% valid condition ($BF_{01} = 3.02$) and the 4 cue–100% condition ($BF_{01} = 3.235$), for which there was moderate evidence of no difference.

We tested whether older adults' higher non-target error rate on 1 cue–50% uncued trials was due to them swapping in the cued item or whether were they equally likely to report the other uncued items. This potential bias towards the cued item can be graphically depicted in a scatter plot of responses by having the cued colour on the x-axis and the reported colour on the y-axis, where responses to the cued item colour should fall on the diagonal. Thus, a linear trend would suggest a tendency to report the cued item in place of the uncued probe, as is apparent in older adults' responses in Figure 3. To quantify this relationship, a linear mixed-effect model was performed separately in young and older adults predicting the response colour from the target (probed, lower priority) colour, and secondly the cued (unprobed, higher priority) colour as fixed effects (random intercepts for participants). Target colour should be predictive of response colour if participants are responding correctly; however, in this case, we were primarily interested in whether the remaining error could be explained by the high-priority cued item. Table 3 shows that in young adults, the target colour is a significant predictor of response colour, but the cued colour is not. In contrast, older adults' responses were predicted by both the target colour and the cued colour, suggesting that they often reported the unprobed, higher-priority item when they were probed to respond with one of the low-priority items.

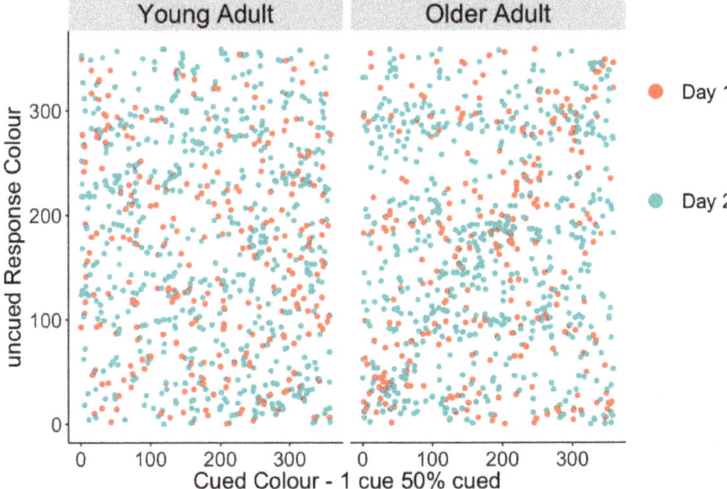

Figure 3. Uncued responses made towards the cued colour for young and older adults. Erroneous responses to the cued item should fall on the diagonal, and a linear trend would suggest a tendency to report the cued item in place of the uncued probe, as is apparent in the older group.

Table 3. Linear mixed-effects model.

Predictors	Young Adult				Older Adult			
	B	CI	*p*	df	*B*	CI	*p*	df
(Intercept)	91.09	77.78–104.40	<0.001	126.16	98.67	85.09–112.25	<0.001	205.80
Target Colour	0.45	0.42–0.49	<0.001	1996.40	0.22	0.17–0.26	<0.001	1996.72
Cued Colour	0.00	−0.04–0.04	0.882	1991.04	0.17	0.13–0.21	<0.001	1996.82
Marginal R^2	0.233				0.067			
Conditional R^2	0.258				0.084			

4. Discussion

The goal of the present study was to determine how age differences in attentional control affect VWM performance when attention is flexibly allocated amongst targets of varying priority. Replicating previous work with younger adults, we show that item priority predicts VWM performance according to a power law, and we extend this finding to older adults. This suggests that despite age-related declines in attentional control, both younger and older adults can flexibly allocate attentional resources according to item priority. Further evidence of this can be taken from our key comparisons: both groups showed equivalent error for items with equivalent probe likelihood and differences in error when the number of cues was the same but probe likelihood differed, suggesting that attention was allocated to items based on priority. A three-part mixture model was used to assess distinct aspects of working memory, revealing that older adults had overall lower precision and higher guess rates. Critically, older adults made more non-target errors when demands on attentional control were highest (i.e., when uncued items in the 1 cue–50% valid condition were tested), and these errors were often the result of swapping in the cued item. This may be because older adults had difficulty inhibiting the more salient (50% likely) item when it was no longer relevant at retrieval (e.g., [62]).

We replicate previous findings in younger adults by showing that the proportion of resources allocated to an item is a useful predictor of VWM performance. A power law where response error was predicted by probe likelihood in a continuous fashion was found to fit the raw error of VWM performance in younger adults and offered a better fit than a simpler linear model. Thus, our data support the long-standing proposal that top–down attentional control and VWM performance are closely intertwined [63,64] and fit with recent models of VWM as a continuous and flexible resource relying on the allocation of spatial attention [6,7].

Importantly, we sought to determine whether this flexible allocation of attention account of working memory performance could similarly explain older adults' response error data. We found that older adults' VWM performance was well fit by a power law where probe likelihood predicts performance. In further support of this model, both older and younger adults reported items of equal priority with equal error (1 cue–50% valid vs. 2 cues–100% valid), and differentiated a single-cued item by its validity (1 cue–100% valid vs. 1 cue–50% valid), suggesting that item priority, rather than the number of cues, determined the allocation of attentional resources. Finally, we also examined older adults' ability to use spatial cues to shift their attention and ignore irrelevant distractors. We compared a no-distractor condition (one item only) to the 1 cue–100% condition, which includes additional distractor items that should be ignored. Previous work has shown that older adults show a greater distractor-related deficit in working memory performance when memory load is high, thus, we may expect both groups to show equivalent performance across conditions wherein the memory load is one item [65]. Indeed, both groups showed no difference between these conditions, suggesting that older adults can use highly predictive cues to direct spatial attention and improve VWM performance [33,66–68]. Taken together, these results suggest that similar to younger adults, older adults allocate attention in a flexible manner according to item priority (albeit with less accuracy), which influences what information is encoded in short-term memory.

Previous work has applied a three-part mixture model of working memory to investigate older adults' VWM in terms of guess rate, precision, and non-target guess rate revealing that age differentially affects these parameters [3,33]. We therefore applied this method to allow for more specific assessments of age-related differences in VWM performance at different attentional priority levels. In terms of precision, we found that older adults have lower precision than younger adults overall and that the fidelity of visual representations in working memory was similarly affected by probe likelihood in both groups [6]. This is in line with previous work showing an age-related decline in the precision of VWM [3].

For guess rate, lower-priority items had a higher guess rate and older adults made more guesses overall, likely reflecting an age-related decrease in working memory resources [30]. Further, the effect of decreasing priority had a steeper effect on older adults' guess rate, who guessed more than younger adults particularly at lower levels of priority. There were no age differences observed in guess rate for the lowest priority condition (1 cue–50% uncued), however, this reflected an increased proportion of non-target errors made by older adults whose cumulative error rate remained significantly higher (Supplemental Figure S2). This likely reflects that as fewer resources are allocated to an item, older adults reach the limit of their available resources [69] and, thus, fail to adequately encode and/or recall some items of lower priority.

Non-target guess rate or swap rate was expected to increase with age, particularly when flexible prioritization demands were high, and we found evidence to this effect. Older adults made more swap errors only when they needed to report an uncued item in the 1 cue–50% condition (i.e., when another item was more highly prioritized). This appeared to be due to older, but not younger, adults reporting the (higher-priority) cued item when the (lower-priority) uncued item was probed. Older adults' tendency to report the salient, but no longer relevant, cued item may reflect their lessened ability to inhibit previously attended information when it becomes task irrelevant, a process that is sometimes referred to as deletion or working memory updating [39,40,62,70,71]. In line with this proposal, older adults have recently been shown to display greater neural activation for no-longer-relevant items during working memory maintenance, an effect that predicts worse recall performance for relevant items [72]. Thus, older adults may have difficulty inhibiting irrelevant items especially when they were once assigned higher priority.

Alternatively, higher reporting of the cued item could reflect a difference in strategy between the groups, such that older adults may have limited their attention to the higher-priority cued item and ignored the lower-priority uncued items (i.e., giving 100% of their attention to the 50% likely item). Refuting this possibility, older adults showed equivalent performance in the 1 cue–50% and 2 cue–100% conditions, when cued items should have received 50% of attentional resources (suggesting that they only gave 50% of their attention to the cued item in the 1 cue–50% condition). Further, if older adults were focusing all their attention on the cued item in the 50% valid condition, we would instead expect performance in this condition to be similar to the other single-cue condition (1 cue–100% valid condition), but error was markedly lower in the 1 cue–100% condition (see Supplementary Figure S1, lower panel). Thus, the swap errors observed in older adults are more likely a product of poor inhibitory control, rather than an ill-adapted strategy.

This work has implications for other populations with poor attentional control, such as children [73] and those diagnosed with ADHD [74], depression [75], and anxiety [76]. For example, deficits in attentional control in ADHD have been linked to deficits in working memory [74]. Recent work investigating VWM in those with ADHD has revealed that performance deficits relate to reduced N2pc amplitude, an EEG component known to be associated with flexible allocation of attention through attentional control [7]. However, these studies have largely focused on memory load and thus, the potential role of attentional control in flexibly allocating attention has been left largely unexplored. Future work may benefit from applying this flexible-attention account of VWM and using mixture models to elucidate the mechanisms through which poor attentional control may influence VWM.

In conclusion, we show that similar to younger adults, older adults' VWM performance relies on the flexible allocation of attention. Older adults show lower precision overall, potentially reflecting lower fidelity of visual representations in working memory and higher guess rates. When demands on flexible resource allocation were highest (on 1 cue–50% uncued item trials), older adults often reported the higher priority (but no longer relevant) cued item, possibly reflecting an age-related inhibitory deficit. The current data elucidates the relationship between flexible allocation of attention models of visual working memory, and the age-related inhibitory deficit by showing that the deletion of no-longer-relevant information is impaired in older adults when attention must be flexibly divided amongst items and task demands are high. Future work should incorporate additional flexible prioritization conditions to further disentangle the relationship between load and probe likelihood and assess potential differences in strategy use between older and younger adults.

Supplementary Materials: The following are available online at http://www.mdpi.com/2076-3425/10/8/542/s1, Figure S1: Key comparison of error for conditions with equal priority and equal cue number, Figure S2: Combined error rate calculated by adding guess rate and non-target error rate for each age group and probe likelihood, Figure S3: Results figure with uncorrected data, Table S1: Summarizing the differences in results with uncorrected data.

Author Contributions: Conceptualization, S.E.H., H.A.L. and E.E.D.; Formal analysis, H.A.L.; Funding acquisition, K.L.C.; Methodology, S.E.H. and E.E.D.; Supervision, S.M.E. and K.L.C.; Writing—original draft, S.E.H. and H.A.L.; Writing—review & editing, E.E.D., S.M.E. and K.L.C. All authors have read and agreed to the published version of the manuscript.

Funding: This research was funded by Natural Sciences and Engineering Research Council of Canada (Grant 2019-435945 to K.L.C. and grant 2019-435945 to S.M.E.) and the Canada Research Chairs program (K.L.C.).

Acknowledgments: We would like to thank Amy Holliday and Uroog Mohammed for their assistance with data collection, and Tyler Kennedy Collins for his assistance with programming.

Conflicts of Interest: The authors declare no conflicts of interest.

References

1. Babcock, R.L.; Salthouse, T.A. Effects of increased processing demands on age differences in working memory. *Psychol. Aging* **1990**, *5*, 421–428. [CrossRef] [PubMed]
2. Gazzaley, A.; Cooney, J.W.; Rissman, J.; D'Esposito, M. Top-down suppression deficit underlies working memory impairment in normal aging. *Nat. Neurosci.* **2005**, *8*, 1298–1300. [CrossRef]

3. Peich, M.-C.; Husain, M.; Bays, P.M. Age-related decline of precision and binding in visual working memory. *Psychol. Aging* **2013**, *28*, 729–743. [CrossRef] [PubMed]
4. Hasher, L.; Zacks, R.T. Working memory, comprehension, and aging: A review and a new view. In *The Psychology of Learning and Motivation*; Academic Press: New York, NY, USA, 1988; Volume 22, pp. 193–225.
5. Sylvain-Roy, S.; Lungu, O.; Belleville, S. Normal aging of the attentional control functions that underlie working memory. *J. Gerontol. B Psychol. Sci. Soc. Sci.* **2015**, *70*, 698–708. [CrossRef] [PubMed]
6. Emrich, S.M.; Lockhart, H.A.; Al-Aidroos, N. Attention mediates the flexible allocation of visual working memory resources. *J. Exp. Psychol. Hum. Percept. Perform.* **2017**, *43*, 1454–1465. [CrossRef]
7. Salahub, C.; Lockhart, H.A.; Dube, B.; Al-Aidroos, N.; Emrich, S.M. Electrophysiological correlates of the flexible allocation of visual working memory resources. *Sci. Rep.* **2019**, *9*, 19428. [CrossRef]
8. Cowan, N. The magical number 4 in short-term memory: A reconsideration of mental storage capacity. *Behav. Brain Sci.* **2001**, *24*, 87–114. [CrossRef]
9. Luck, S.J.; Vogel, E.K. Visual working memory capacity: From psychophysics and neurobiology to individual differences. *Trends Cogn. Sci.* **2013**, *17*, 391–400. [CrossRef]
10. Ma, W.J.; Husain, M.; Bays, P.M. Changing concepts of working memory. *Nat. Neurosci.* **2014**, *17*, 347–356. [CrossRef]
11. Bays, P.M.; Husain, M. Dynamic shifts of limited working memory resources in human vision. *Science* **2008**, *321*, 851–854. [CrossRef]
12. Zhang, W.; Luck, S.J. Discrete fixed-resolution representations in visual working memory. *Nature* **2008**, *453*, 233–235. [CrossRef] [PubMed]
13. Bays, P.M.; Catalao, R.F.G.; Husain, M. The precision of visual working memory is set by allocation of a shared resource. *J. Vis.* **2009**, *9*, 7. [CrossRef] [PubMed]
14. Van den Berg, R.; Shin, H.; Chou, W.-C.; George, R.; Ma, W.J. Variability in encoding precision accounts for visual short-term memory limitations. *Proc. Natl. Acad. Sci. USA* **2012**, *109*, 8780–8785. [CrossRef] [PubMed]
15. Klyszejko, Z.; Rahmati, M.; Curtis, C.E. Attentional priority determines working memory precision. *Vis. Res.* **2014**, *105*, 70–76. [CrossRef] [PubMed]
16. Machizawa, M.G.; Goh, C.C.W.; Driver, J. Human visual short-term memory precision can be varied at will when the number of retained items is low. *Psychol. Sci.* **2012**, *23*, 554–559. [CrossRef]
17. Dube, B.; Emrich, S.M.; Al-Aidroos, N. More than a filter: Feature-based attention regulates the distribution of visual working memory resources. *J. Exp. Psychol. Hum. Percept. Perform.* **2017**, *43*, 1843–1854. [CrossRef]
18. Adam, K.C.S.; Mance, I.; Fukuda, K.; Vogel, E.K. The Contribution of Attentional Lapses to Individual Differences in Visual Working Memory Capacity. *J. Cogn. Neurosci.* **2015**, *27*, 1601–1616. [CrossRef]
19. Awh, E.; Vogel, E.K.; Oh, S.-H. Interactions between attention and working memory. *Neuroscience* **2006**, *139*, 201–208. [CrossRef]
20. Engle, R.W. Working memory capacity as executive attention. *Curr. Dir. Psychol. Sci.* **2002**, *11*, 19–23. [CrossRef]
21. Fukuda, K.; Woodman, G.F.; Vogel, E.K. Individual Differences in Visual Working Memory Capacity. In *Mechanisms of Sensory Working Memory*; Elsevier: Amsterdam, The Netherlands, 2015; pp. 105–119. ISBN 978-0-12-801371-7.
22. Cowan, N.; Morey, C.C. Visual working memory depends on attentional filtering. *Trends Cogn. Sci.* **2006**, *10*, 139–141. [CrossRef]
23. McCollough, A.W.; Machizawa, M.G.; Vogel, E.K. Electrophysiological measures of maintaining representations in visual working memory. *Cortex* **2007**, *43*, 77–94. [CrossRef]
24. Bays, P.M. Reassessing the evidence for capacity limits in neural signals related to working memory. *Cereb. Cortex* **2018**, *28*, 1432–1438. [CrossRef] [PubMed]
25. Vogel, E.K.; McCollough, A.W.; Machizawa, M.G. Neural measures reveal individual differences in controlling access to working memory. *Nature* **2005**, *438*, 500–503. [CrossRef] [PubMed]
26. Fukuda, K.; Vogel, E.K. Human Variation in Overriding Attentional Capture. *J. Neurosci.* **2009**, *29*, 8726–8733. [CrossRef] [PubMed]
27. Emrich, S.M.; Busseri, M.A. Re-evaluating the relationships among filtering activity, unnecessary storage, and visual working memory capacity. *Cogn. Affect. Behav. Neurosci.* **2015**, *15*, 589–597. [CrossRef] [PubMed]
28. Eimer, M. The N2pc component as an indicator of attentional selectivity. *Electroencephalogr. Clin. Neurophysiol.* **1996**, *99*, 225–234. [CrossRef]

29. Hickey, C.; McDonald, J.J.; Theeuwes, J. Electrophysiological evidence of the capture of visual attention. *J. Cogn. Neurosci.* **2006**, *18*, 604–613. [CrossRef]
30. Brockmole, J.R.; Logie, R.H. Age-related change in visual working memory: A study of 55,753 participants aged 8–75. *Front. Psychol.* **2013**, *4*, 12. [CrossRef]
31. Jost, K.; Bryck, R.L.; Vogel, E.K.; Mayr, U. Are old adults just like low working memory young adults? Filtering efficiency and age differences in visual working memory. *Cereb. Cortex* **2011**, *21*, 1147–1154. [CrossRef]
32. Reuter-Lorenz, P.A.; Sylvester, C.C. The cognitive neuroscience of working memory and aging. In *Cognitive Neuroscience of Aging: Linking Cognitive and Cerebral Aging*; Oxford University Press: New York, NY, USA, 2005; pp. 186–217.
33. Mok, R.M.; Myers, N.E.; Wallis, G.; Nobre, A.C. Behavioral and neural markers of flexible attention over working memory in aging. *Cereb. Cortex* **2016**, *26*, 1831–1842. [CrossRef]
34. Gazzaley, A.; Clapp, W.; Kelley, J.; McEvoy, K.; Knight, R.T.; D'Esposito, M. Age-related top-down suppression deficit in the early stages of cortical visual memory processing. *Proc. Natl. Acad. Sci. USA* **2008**, *105*, 13122–13126. [CrossRef] [PubMed]
35. Schwarzkopp, T.; Mayr, U.; Jost, K. Early selection versus late correction: Age-related differences in controlling working memory contents. *Psychol. Aging* **2016**, *31*, 430–441. [CrossRef] [PubMed]
36. Cashdollar, N.; Fukuda, K.; Bocklage, A.; Aurtenetxe, S.; Vogel, E.K.; Gazzaley, A. Prolonged disengagement from attentional capture in normal aging. *Psychol. Aging* **2013**, *28*, 77–86. [CrossRef] [PubMed]
37. Kato, K.; Nakamura, A.; Kato, T.; Kuratsubo, I.; Yamagishi, M.; Iwata, K.; Ito, K. Age-related changes in attentional control using an n-back working memory paradigm. *Exp. Aging Res.* **2016**, *42*, 390–402. [CrossRef]
38. Hasher, L.; Zacks, R.T.; May, C.P. Inhibitory control, circadian arousal, and age. In *Attention and Performance*; MIT Press: Cambridge, MA, USA, 1999; pp. 653–675.
39. Lustig, C.; Hasher, L.; Zacks, R.T. Inhibitory deficit theory: Recent developments in a "new view". In *Inhibition in Cognition*; Gorfein, D.S., MacLeod, C.M., Eds.; American Psychological Association: Washington, DC, USA, 2007; pp. 145–162. ISBN 978-1-59147-930-7.
40. Persad, C.C.; Abeles, N.; Zacks, R.T.; Denburg, N.L. Inhibitory changes after age 60 and their relationship to measures of attention and memory. *J. Gerontol. B. Psychol. Sci. Soc. Sci.* **2002**, *57*, P223–P232. [CrossRef]
41. Brockmole, J.R.; Parra, M.A.; Sala, S.D.; Logie, R.H. Do binding deficits account for age-related decline in visual working memory? *Psychon. Bull. Rev.* **2008**, *15*, 543–547. [CrossRef]
42. Cowan, N.; Naveh-Benjamin, M.; Kilb, A.; Saults, J.S. Life-span development of visual working memory: When is feature binding difficult? *Dev. Psychol.* **2006**, *42*, 1089–1102. [CrossRef]
43. Parra, M.A.; Abrahams, S.; Logie, R.H.; Sala, S.D. Age and binding within-dimension features in visual short-term memory. *Neurosci. Lett.* **2009**, *449*, 1–5. [CrossRef]
44. Sander, M.C.; Werkle-Bergner, M.; Lindenberger, U. Binding and strategic selection in working memory: A lifespan dissociation. *Psychol. Aging* **2011**, *26*, 612–624. [CrossRef]
45. Mitchell, D.J.; Cusack, R. Visual short-term memory through the lifespan: Preserved benefits of context and metacognition. *Psychol. Aging* **2018**, *33*, 841–854. [CrossRef]
46. Carson, N.; Leach, L.; Murphy, K.J. A re-examination of Montreal Cognitive Assessment (MoCA) cutoff scores: Re-examination of MoCA cutoff scores. *Int. J. Geriatr. Psychiatry* **2018**, *33*, 379–388. [CrossRef] [PubMed]
47. Nasreddine, Z.S.; Phillips, N.A.; Bédirian, V.; Charbonneau, S.; Whitehead, V.; Collin, I.; Cummings, J.L.; Chertkow, H. The Montreal Cognitive Assessment, MoCA: A brief screening tool for Mild Cognitive Impairment. *J. Am. Geriatr. Soc.* **2005**, *53*, 695–699. [CrossRef] [PubMed]
48. Verhaeghen, P. Aging and vocabulary score: A meta-analysis. *Psychol. Aging* **2003**, *18*, 322–339. [CrossRef] [PubMed]
49. Peirce, J.; Gray, J.R.; Simpson, S.; MacAskill, M.; Höchenberger, R.; Sogo, H.; Kastman, E.; Lindeløv, J.K. PsychoPy2: Experiments in behavior made easy. *Behav. Res. Methods* **2019**, *51*, 195–203. [CrossRef] [PubMed]
50. Schurgin, M.W.; Wixted, J.T.; Brady, T.F. Psychophysical scaling reveals a unified theory of visual memory strength. *bioRxiv* **2018**, 325472.
51. Canty, A.; Ripley, B.D. *Boot: Bootstrap R (S-Plus) Functions*; R package version 1.3-25. 2020.
52. Davison, A.C.; Hinkley, D.V. *Bootstrap Methods and Their Applications*; Cambridge University Press: Cambridge, UK, 1997.

53. Bates, D.; Mächler, M.; Bolker, B.; Walker, S. Fitting linear mixed-effects models using lme4. *J. Stat. Softw.* **2015**, *67*. [CrossRef]
54. Kuznetsova, A.; Brockhoff, P.B.; Christensen, R.H.B. lmerTest package: Tests in linear mixed effects models. *J. Stat. Softw.* **2017**, *82*, 1–26. [CrossRef]
55. Fox, J.; Weisberg, S. *An R Companion to Applied Regression*, 3rd ed.; Sage: Thousand Oaks, CA, USA, 2019.
56. Lüdecke, D. sjPlot: Data Visualization for Statistics in Social Science. 2020. Available online: https://CRAN.R-project.org/package=sjPlot (accessed on 30 June 2020).
57. Kenward, M.G.; Roger, J.H. Small sample inference for fixed effects from restricted maximum likelihood. *Biometrics* **1997**, *53*, 983. [CrossRef]
58. JASP Team. JASP. Available online: https://jasp-stats.org/ (accessed on 30 June 2020).
59. Van Doorn, J.; van den Bergh, D.; Bohm, U.; Dablander, F.; Derks, K.; Draws, T.; Etz, A.; Evans, N.J.; Gronau, Q.F.; Hinne, M.; et al. The JASP Guidelines for Conducting and Reporting a Bayesian Analysis. *PsyArXiv* **2019**. [CrossRef]
60. Suchow, J.W.; Brady, T.F.; Fougnie, D.; Alvarez, G.A. Modeling visual working memory with the Mem Toolbox. *J. Vis.* **2013**, *13*, 9. [CrossRef]
61. Allen, M.; Poggiali, D.; Whitaker, K.; Marshall, T.R.; Kievit, R.A. Raincloud plots: A multi-platform tool for robust data visualization. *Wellcome Open Res.* **2019**, *4*, 63. [CrossRef] [PubMed]
62. Healey, M.K.; Hasher, L.; Campbell, K.L. The role of suppression in resolving interference: Evidence for an age-related deficit. *Psychol. Aging* **2013**, *28*, 721–728. [CrossRef] [PubMed]
63. Awh, E.; Jonides, J. Overlapping mechanisms of attention and spatial working memory. *Trends Cogn. Sci.* **2001**, *5*, 119–126. [CrossRef]
64. Gazzaley, A.; Nobre, A.C. Top-down modulation: Bridging selective attention and working memory. *Trends Cogn. Sci.* **2012**, *16*, 129–135. [CrossRef]
65. Gazzaley, A.; Sheridan, M.A.; Cooney, J.W.; D'Esposito, M. Age-related deficits in component processes of working memory. *Neuropsychology* **2007**, *21*, 532–539. [CrossRef]
66. Loaiza, V.M.; Souza, A.S. Is refreshing in working memory impaired in older age? Evidence from the retro-cue paradigm: Refreshing in aging. *Ann. N. Y. Acad. Sci.* **2018**, *1424*, 175–189. [CrossRef]
67. Souza, A.S. No age deficits in the ability to use attention to improve visual working memory. *Psychol. Aging* **2016**, *31*, 456–470. [CrossRef]
68. Tales, A.; Muir, J.L.; Bayer, A.; Snowden, R.J. Spatial shifts in visual attention in normal ageing and dementia of the Alzheimer type. *Neuropsychologia* **2002**, *40*, 2000–2012. [CrossRef]
69. Reuter-Lorenz, P.A.; Cappell, K.A. Neurocognitive aging and the compensation hypothesis. *Curr. Dir. Psychol. Sci.* **2008**, *17*, 177–182. [CrossRef]
70. Cansino, S.; Guzzon, D.; Martinelli, M.; Barollo, M.; Casco, C. Effects of aging on interference control in selective attention and working memory. *Mem. Cognit.* **2011**, *39*, 1409–1422. [CrossRef]
71. Miyake, A.; Friedman, N.P.; Emerson, M.J.; Witzki, A.H.; Howerter, A.; Wager, T.D. The unity and diversity of executive functions and their contributions to complex "frontal lobe" tasks: A latent variable analysis. *Cognit. Psychol.* **2000**, *41*, 49–100. [CrossRef]
72. Weeks, J.C.; Grady, C.L.; Hasher, L.; Buchsbaum, B.R. Holding on to the past: Older adults Show lingering neural activation of no-longer-relevant items in working memory. *J. Cogn. Neurosci.* **2020**, 1–17. [CrossRef]
73. Davidson, M.C.; Amso, D.; Anderson, L.C.; Diamond, A. Development of cognitive control and executive functions from 4 to 13 years: Evidence from manipulations of memory, inhibition, and task switching. *Neuropsychologia* **2006**, *44*, 2037–2078. [CrossRef] [PubMed]
74. Alderson, R.M.; Kasper, L.J.; Hudec, K.L.; Patros, C.H.G. Attention-deficit/hyperactivity disorder (ADHD) and working memory in adults: A meta-analytic review. *Neuropsychology* **2013**, *27*, 287–302. [CrossRef] [PubMed]

75. Silton, R.L.; Heller, W.; Engels, A.S.; Towers, D.N.; Spielberg, J.M.; Edgar, J.C.; Sass, S.M.; Stewart, J.L.; Sutton, B.P.; Banich, M.T.; et al. Depression and anxious apprehension distinguish frontocingulate cortical activity during top-down attentional control. *J. Abnorm. Psychol.* **2011**, *120*, 272–285. [CrossRef]
76. Salahub, C.; Emrich, S. Fear not! Anxiety biases attentional enhancement of threat without impairing working memory filtering. *PsyArXiv* **2019**. [CrossRef]

Article

Visuo-Perceptual and Decision-Making Contributions to Visual Hallucinations in Mild Cognitive Impairment in Lewy Body Disease: Insights from a Drift Diffusion Analysis

Lauren Revie [1,*], Calum A Hamilton [2], Joanna Ciafone [2], Paul C Donaghy [2], Alan Thomas [2] and Claudia Metzler-Baddeley [1]

1 Cardiff University Brain Research Imaging Centre (CUBRIC), School of Psychology, Cardiff University, Maindy Road, Cardiff CF24 4HQ, UK; metzler-baddeleyc@cardiff.ac.uk
2 Translational and Clinical Research Institute, Newcastle University Biomedical Research Building, Campus for Ageing and Vitality, Newcastle Upon Tyne NE4 5PL, UK; c.hamilton3@newcastle.ac.uk (C.A.H.); joanna.ciafone@newcastle.ac.uk (J.C.); paul.donaghy@newcastle.ac.uk (P.C.D.); alan.thomas@newcastle.ac.uk (A.T.)
* Correspondence: reviel@cardiff.ac.uk

Received: 30 June 2020; Accepted: 7 August 2020; Published: 11 August 2020

Abstract: *Background:* Visual hallucinations (VH) are a common symptom in dementia with Lewy bodies (DLB); however, their cognitive underpinnings remain unclear. Hallucinations have been related to cognitive slowing in DLB and may arise due to impaired sensory input, dysregulation in top-down influences over perception, or an imbalance between the two, resulting in false visual inferences. *Methods:* Here we employed a drift diffusion model yielding estimates of perceptual encoding time, decision threshold, and drift rate of evidence accumulation to (i) investigate the nature of DLB-related slowing of responses and (ii) their relationship to visuospatial performance and visual hallucinations. The EZ drift diffusion model was fitted to mean reaction time (RT), accuracy and RT variance from two-choice reaction time (CRT) tasks and data were compared between groups of mild cognitive impairment (MCI-LB) LB patients (n = 49) and healthy older adults (n = 25). *Results:* No difference was detected in drift rate between patients and controls, but MCI-LB patients showed slower non-decision times and boundary separation values than control participants. Furthermore, non-decision time was negatively correlated with visuospatial performance in MCI-LB, and score on visual hallucinations inventory. However, only boundary separation was related to clinical incidence of visual hallucinations. *Conclusions:* These results suggest that a primary impairment in perceptual encoding may contribute to the visuospatial performance, however a more cautious response strategy may be related to visual hallucinations in Lewy body disease. Interestingly, MCI-LB patients showed no impairment in information processing ability, suggesting that, when perceptual encoding was successful, patients were able to normally process information, potentially explaining the variability of hallucination incidence.

Keywords: Lewy body disease; mild cognitive impairment; attention; visual hallucinations; perception; drift diffusion; vision

1. Introduction

Dementia with Lewy bodies (DLB) is clinically characterized by four core clinical features, frequent cognitive fluctuations, complex visual hallucinations, rapid eye movement sleep behavior disorder (RBD), and Parkinsonism [1]. Cognitively, DLB patients also experience disproportionate impairment in attention and visual perceptual abilities in comparison to both older adults and patients with

Alzheimer's disease (AD) [1–4]. The question arises how these cognitive and perceptual impairments relate to clinical symptoms in DLB. For instance, visual hallucinations may arise from bottom up visuoperceptual and top-down attentional impairments [5], and similarly, cognitive fluctuations may be related to attentional deficits [6].

Mild cognitive impairment with Lewy bodies (MCI-LB) refers to individuals experiencing mild cognitive impairment which is likely to progress to DLB [7]. MCI refers to the stage between normal aging and dementia, in which cognitive decline occurs, but activities of daily living are relatively preserved [8]. Patients may be diagnosed with possible MCI-LB if one core clinical feature or biomarker is present, or patients may be diagnosed with probable MCI-LB if two or more clinical features are present, or one clinical feature plus one biomarker according to the proposed MCI-LB criteria [7]. Cognitively, MCI-LB patients experience marked impairments in processing speed, executive dysfunction and visuospatial functions [3]. A greater slowing of response times (RT) in DLB compared with AD patients and healthy older controls, particularly in tasks with increasing complexity of cognitive processing has been frequently reported (i.e., in choice reaction task; CRT) [9,10]. For instance, DLB patients show disproportional slowing in flanker performance, but can exhibit the flanker effect when alerted to the trial [11]. Moreover, DLB patients exhibit large intraindividual fluctuation in RT performance, that may underpin cognitive fluctuation [10,12]. However, not all studies have found variability in CRT performance to be related to visual or attentional performance, thus this link remains unclear [13]. Currently, the nature of RT slowing and variability in DLB and their relationship to clinical symptoms is not well understood. To address this, the present study adopted the diffusion drift model (DDM) to gain a better understanding of the nature of RT slowing in DLB patients.

Based on the assumptions that decisions occur over time and are prone to errors, the DDM can be applied to CRTs. The drift diffusion model [14,15] is a sequential-sampling model, which assumes that information driving a decision is accumulated over time, until it reaches one of two response boundaries (denoted by "upper" and "lower" response boundaries in Figure 1) and forms the ultimate response (for example, a "left" or "right" response would be the upper or lower response boundaries in a model as depicted in Figure 1). According to the diffusion model, overall processing is segmented into several components that contribute to ultimate performance: (1) first, components of processing that do not include active decision-making, i.e., such as perceptual encoding and response execution which are represented by the "non-decision time (t)", (2) second, a criterion threshold that information accumulates to, which is represented by the "boundary separation (a)", and finally, (3) the rate that information accumulates towards a decision, estimated by the "drift rate (v)". The DDM assumes that the accumulation of information (drift rate) during a decision process is not constant but varies over time, represented by (s). This variability also contributes to error responses, where the accumulation of information can reach the incorrect boundary. For example, in the context of a flanker task in which stimuli are either congruent and thus "easy", or incongruent and thus "harder", drift rates will tend to be more prone to variation in harder conditions, and therefore response times are slower and have a higher probability of reaching an incorrect response boundary.

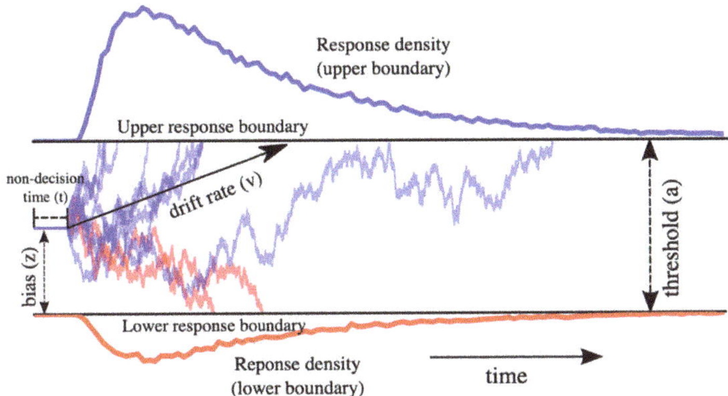

Figure 1. The drift diffusion model. Red and blue denote the information being accumulated during one trial towards one of two responses in a 2-choice task (for example, left or right). Lines in the center of the figure represent the noisy accumulation of evidence for either decision during a trial. In this instance, more evidence is accumulated toward a blue (upper) response. Reaction times from each trial are fit to this model to return an average estimate of non-decision time (t), boundary separation or threshold (a), and drift rate (v). Adapted from Wiecki et al. [16].

The DDM has been extensively applied in aging research due to the effect of aging on the speed accuracy trade off. Most consistent findings from drift diffusion modelling in older adults show increased boundary separation values and lengthening of non-decision times despite no difference in accuracy between older and younger adults [17,18]. Drift rate remains relatively preserved in older adults, indicating a shift in response strategy to a more cautious decision-making approach.

The DDM was applied to CRT performance of Parkinson's (PD) patients with hallucinations relative to PD without hallucinations and healthy controls. Slower drift rates were found in PD patients with hallucinations, and shorter perceptual encoding times were found in all PD patients compared to controls, suggesting that the accumulation of evidence was hindered by perceptual encoding problems [19]. The authors proposed that impaired sensory evidence accumulation may lead to reduced information processing quality and an over-reliance on top down processing in PD, which in turn may underpin visual hallucinations.

The question arises whether DLB patients share similar impairments in perceptual and evidence accumulation as PD patients and whether these can be detected at the MCI-LB stage. By employing a variation of the DDM—the EZ model [20] to assess elements of CRT that may indicate a primary perceptual or attentional impairment, we may be able to narrow down those cognitive processes that underpin visual impairments, cognitive slowing and visual hallucinations in MCI-LB patients. This is of importance, as hallucinations in Lewy body disorders have been proposed to be the product of impaired sensory input and dysfunctional attentional in terms of faulty top-down attentional control [21,22], but the relative contribution of these processes remains unclear. Moreover, visual hallucinations are a predictor of cognitive decline in PD and have a significant effect on the quality of life and treatment of patients and caregivers in both PD and DLB [23], therefore there is a clear clinical need in understanding this symptom. As both DLB and PD patients often experience visual hallucinations, and share pathological features, it was hypothesized that MCI-LB patients would show similar impairments in drift rate and perceptual encoding times as those previously reported for PD patients with hallucinations.

2. Materials and Methods

2.1. Participants

Patients were part of the Lewy-Pro cohort [24]. Participants have been described in detail previously [24,25]. Briefly, individuals were recruited from memory clinics in the North East of England and Cumbria, were over 60 years of age and met NIA-AA criteria for MCI. All participants were clinically assessed for core features of DLB by an experienced psychiatrist (PD). Additional assessments included neuropsychological assessment and dopaminergic imaging using 123I-N-fluoropropyl-2β-carbomethoxy-3β-(4-iodophenyl) single-photon emission computed tomography (FP-CIT SPECT), which were rated normal/abnormal blind to clinical information by an expert panel.

Diagnosis was made by a consensus panel of old age psychiatrists, and participants were categorized into one of two groups; probable MCI-LB in which patients had two or more of the core diagnostic features of DLB or one feature and an abnormal FP-CIT scan, and possible MCI-DLB in which patients had one core feature or an abnormal FP-CIT scan. These diagnoses match the categories of probable MCI-LB and possible MCI-LB in the recently published international consensus criteria [7]. Patients were excluded if they had dementia, an MMSE score <20, a CDR score of >0.5, Parkinsonism which developed more than one year prior to cognitive impairment, or evidence of stroke, neurological or medical conditions which would affect performance in assessments. For this investigation, data from a subset of 50 participants described previously by Donaghy et al., [24] were accessed, including CRT data, and clinical assessments. One participant (probable MCI-LB) was excluded due to limited CRT data.

Healthy controls were recruited from the School of Psychology volunteer's panel at Cardiff University, local age interest groups and the Join Dementia Research community in Cardiff, South Wales, UK. Controls had no visual disturbances, no history of psychiatric illness, no current diagnosis of dementia or cognitive impairment, and were over the age of 60 years. A group of younger control participants was also included in the analysis, which allowed us to contrast disease and normal aging-related differences in the drift diffusion parameters. These participants were recruited from a School of Psychology undergraduate student database at Cardiff University, and were subject to the same exclusion criteria as the older control participants. Table 1 shows demographic information of the cohort.

Table 1. Demographic information of Lewy-Pro cohort and control participants. ACE = Addenbrooke's Cognitive Examination, NEVHI = North East Visual Hallucinations Inventory, UPDRS = Unified Parkinson's Disease Rating Scale, CAF = Clinical Assessment of Fluctuations.

	Probable MCI- LB (*n* = 37)	Possible MCI- LB (*n* = 12)	Older Controls (*n* = 25)	Younger Controls (*n* = 25)
Age	72.86 (15.51)	75.25 (7.3)	68.36 (6.11)	21.69 (2.85)
Sex	Female = 13	Female = 3	Female = 13	Female = 15
Education	11.57 (2.85)	10.75 (2.09)	15.12 (2.40)	14.95 (1.89)
ACE	78.34 (8.88)	79.33 (14.09)	93.52 (4.18)	92.38 (3.12)
Visual Hallucinations	20	2	-	-
NEVHI	3.66 (4.47)	1 (3.316)	0	0
UPDRS	25.21 (15.71)	15.36 (7.75)	-	-
CAF	2.70 (3.03)	2.16 (2.48)	0.32 (0.63)	0
Cholinesterase Inhibitor	18	0	-	-
Levodopa	8	0	-	-

2.2. Materials

Lewy-Pro participants' general cognitive performance was assessed with the Addenbrooke's Cognitive Exam (ACE-R; [26], verbal fluency with the FAS Verbal Fluency and Graded Naming Tests [27], attention-switching with the Trail-making test parts A and B and verbal memory with the Rey Auditory Verbal learning test (RAVLT) [28]. Participants completed computerized assessments of

CRT, digit vigilance and visuoperceptual functions, in addition to clinical assessments including the Geriatric Depression Scale (GDS) [29], Clinician Assessment of Fluctuations [30], Dementia Cognitive Fluctuations Scale (DCFS) [31], Neuropsychiatric Inventory (NPI) [32], North East Visual Hallucinations Interview (NEVHI) [33], Unified Parkinson's Disease Rating Scale (UPDRS) [34], and the Instrumental Activities of Daily Living scale (ADL) [35]. Control participants were also assessed using a computerized assessment of choice reaction time, the ACE-R, CAF and NEVHI.

2.3. Procedure

MCI-LB participants completed a simple CRT 'left or right' arrow response task (30 trials) as described in Donaghy et al. [24]. RT data for each control participant were collected using a CRT flanker task [36]. The task consisted of five horizontal arrows presented on the screen. Participants were instructed to attend to the central arrow and respond to the direction the target arrow pointed to, using either left or right arrow keys. Central arrows were flanked either by lines (neutral condition) or by arrows that either pointed in the same direction as the central arrow (congruent condition) or in its opposite direction (incongruent condition). Targets were presented on the screen for 1700 ms, and participants were asked to respond as quickly and as accurately as possible. Participants completed 96 trials in each block, for a duration of 5 blocks, with a rest of 30 s between each block, totaling 480 trials. For the purposes of the DDM analyses and to account for differences in complexity of tasks between groups, EZ model parameters for the control participants were calculated only from the RT data of the "neutral" condition in the CRT flanker task (number trials =160).

The means of RT, accuracy and RT variance were generated for each participant across all completed trials. Participants were excluded from the analysis if they did not complete the task ($n = 1$). A simplified version of the original drift diffusion model, the EZ DDM [20] was then used to estimate each participants' drift rate [v], boundary separation [a], and non-decision time [t] from mean and variance of RTs in correct decision trials and from accuracy (proportion correct). The EZ model outperforms other DDM in fitting CRT data from limited trial numbers (<100), and thus is particularly suitable for the modelling of data from patients with reduced cognitive abilities [37]. The EZ model has also been shown to be sensitive to responses which may arise from sources outside the diffusion process assumptions [38], although also see Ratcliff et al. for further information [39].

2.4. Statistical Analyses

Group differences in mean RT, accuracy, RT variance, and the three EZ DDM parameters (non-decision time, boundary separation, and drift rate) were tested between young and older participants to assess age-related effects, and between older participants and possible MCI-LB and probable MCI-LB groups respectively, to assess disease-related effects. Variance was assessed using Levene's test for homogeneity of variance. Due to the high variance in patient data, non-parametric Kruskall–Wallis tests, with post-hoc Mann–Whitney U tests were employed for all group comparisons. All multiplicity of post-hoc testing was corrected with 5% False Discover Rate (FDR) using the Benjamini–Hochberg procedure [40] where p_{cor} is reported as FDR corrected p-values.

Following this, linear hierarchical stepwise regression models were fit to demographic and DDM data in order to determine the best predictors of clinical measures in both possible and probable MCI-LB groups. Spearman rho correlations were calculated between significant predictors and clinical measures to assess the direction of effects.

Patient groups were also categorized on the basis of their NEVHI score, with a high NEVHI indicating greater incidence of visual disturbances. This allowed us to compare drift parameters between those patients who experienced symptoms without visual hallucinations, (NEVHI score = 0; NEVHI−, $n = 29$) and those who experienced symptoms with visual hallucinations (NEVHI score > 1; NEVHI+, $n = 20$), see [19]. In addition, these comparisons were also conducted between patients with clinically rated absence ($n = 37$) or presence ($n = 12$) of complex visual hallucinations. Mann–Whitney U tests were used to assess group differences in DDM parameters.

3. Results

3.1. Demographics

Kruskal–Wallis H tests revealed group differences in age ($\chi 2$ (2) = 12.558, p = 0.002) and education ($\chi 2$ (2) = 22.280, <0.001). Both probable and possible MCI-LB groups were significantly older than healthy control participants (probable: U = 189, p_{cor} < 0.001; possible: U = 65, p_{cor} = 0.02). In addition, control participants had significantly more years of education than possible (U = 241.5, p_{cor} < 0.001) and probable MCI-LB groups (U = 661.5, p_{cor} = 0.001).

3.2. Mean RT, Accuracy and Variance Differences between Older Versys Younger Controls and between Older Controls and MCI-LB Groups

Kruskal–Wallis H tests showed group differences in mean RT ($\chi 2$ (2) = 15.747, p < 0.001), mean accuracy ($\chi 2$ (2) = 16.927, p < 0.001), mean RT variance ($\chi 2$ (2) = 7.589, p = 0.022) and standard deviation in RT ($\chi 2$ (2) = 26.503, p < 0.001). Younger participants had significantly lower RTs than older controls (U = 499, p_{cor} < 0.001) (Figure 2A); however, younger and older adults did not significantly differ in accuracy (p = 0.17) (Figure 2B). Both RT variance (p = 0.25) and RT standard deviation (p = 0.621) did not differ between younger and older adults.

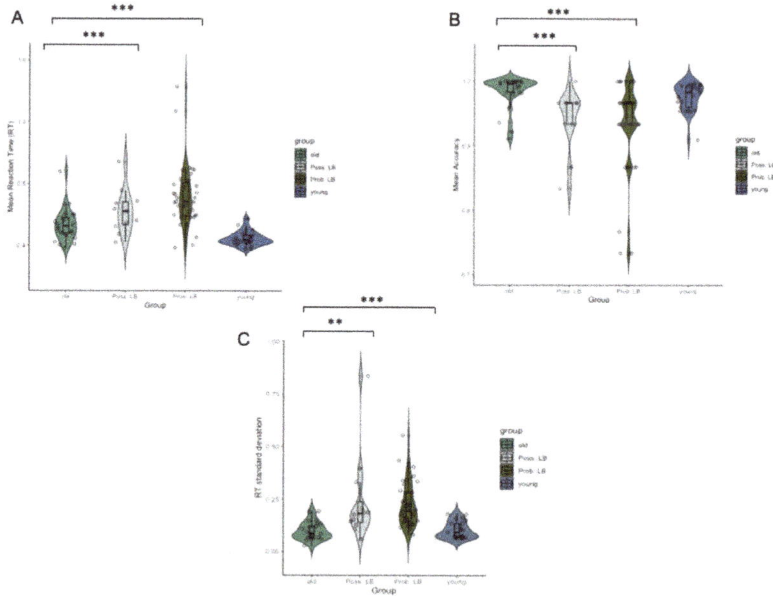

Figure 2. (**A**) Diagnostic group differences in mean reaction time (RT) (**B**) Group differences in mean Old = older healthy controls, young = younger healthy controls, Poss. LB = possible MCI-LB, Prob. LB = probable MCI-LB. (**C**) Standard deviation of RT is significantly greater in possible and probable MCI-LB than older adults. ** p_{cor} = 0.01, *** p_{cor} < 0.001.

Mean RTs were significantly lower in the older control relative to the probable MCI-LB group (U = 157, p_{cor} < 0.001) (Figure 2A). Healthy older controls also had higher accuracy scores than possible (U = 46, p_{cor} < 0.001) and probable MCI-LB patients (U = 215, p_{cor} < 0.001) (Figure 2B). Reaction time variance did not differ between healthy older controls and possible MCI-LB (p_{cor} = 0.82), or probable MCI-LB (p_{cor} = 0.64). Standard deviation was significantly greater in possible MCI-LB (p_{cor} = 0.0015) and probable MCI-LB (p_{cor} < 0.001) in comparison to healthy older controls.

3.3. Group Differences in DDM Parameters

Kruskal–Wallis H tests showed significant differences between all groups in non-decision time ($\chi 2$ (2) = 20.429, $p < 0.001$) and boundary separation values ($\chi 2$ (2) = 9.808, $p = 0.007$), but not drift rates ($\chi 2$ (2) = 0.174, $p = 0.917$). Post-hoc Mann–Whitney U tests revealed that older adults had significantly longer boundary separation values than younger adults (U = 415, $p_{cor} = 0.04$). No significant difference in non-decision time ($p = 0.11$) between older and younger adults were observed.

Post-hoc Mann–Whitney U tests showed significantly lower boundary separation values in both possible MCI-LB (U = 53, $p_{cor} = 0.004$), and probable MCI-LB (U = 240, $p_{cor} = 0.009$) in comparison to older adults. Moreover, greater non-decision times were observed in possible MCI-LB (U = 56, $p_{cor} = 0.005$), and probable MCI-LB (U = 125, $p_{cor} \leq 0$ 0.001) in comparison to older adults. In addition, no significant differences in boundary separation value ($p_{cor} = 0.61$) and non-decision time ($p_{cor} = 0.40$) were observed between possible and probable MCI-LB. Mann–Whitney U tests also showed significant differences in variances between drift rate, non-decision time and boundary separation values between older adults, possible MCI-LB and probable MCI-LB groups (drift: U = 23.5, $p_{cor} < 0.001$; non-decision: U = 444, $p_{cor} \leq 0.001$; boundary: U = 420, $p_{cor} < 0.001$). Additionally, Levene's test of variance showed significantly greater variation in non-decision time (F (2, 69) = 8.41, $p < 0.001$), boundary separation (F (2, 68) = 3.26, $p = 0.041$), and drift rate (F (2, 71) = 9.809, $p < 0.001$) between control participants and the probable MCI-LB group (Figure 3).

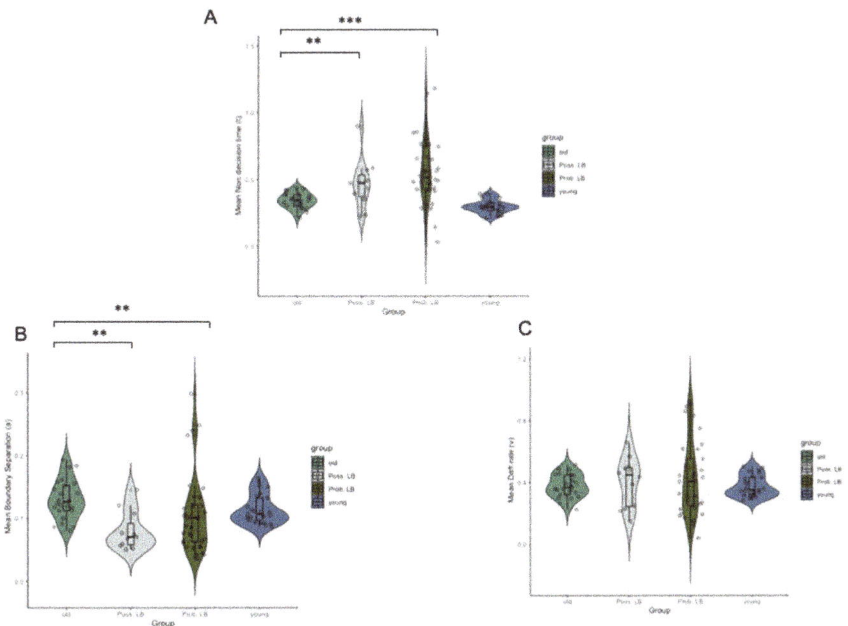

Figure 3. Diagnostic group differences in (**A**) non-decision time (t) and (**B**) boundary separation (a) and (**C**) drift rate DDM values. ** $p_{cor} = 0.01$, *** $p_{cor} \leq 0.001$.

3.4. Effect of Age and Education on Group Differences

The potential influence of age and education on group differences was assessed given that the participants with possible and probable MCI-LB were older and less educated compared to the older control group. This was done by repeating the patient-control comparisons for a subset of control participants aged over 70 ($n = 11$), in which age ($\chi 2$ (2) = 1.387, $p = 0.50$) and education ($\chi 2$ (2) = 5.980, $p = 0.06$) did not significantly differ from the patient groups. The same pattern of results was observed for these sub-group comparisons, in which mean RT was significantly lower ($\chi2$ (2) = 11.059, p = 0.004) and mean accuracy significantly higher in older adults ($\chi2$ (2) = 12.139, p = 0.002) than both patient groups. Similarly, mean non-decision time was significantly lower in older adults ($\chi2$ (2) = 12.844, p = 0.002) and boundary separation greater in older adults ($\chi2$ (2) = 6.253, p = 0.044), with no significant difference in drift rate ($\chi2$ (2) = 0.082, p = 0.960). Thus, age and education did not appear to significantly affect the differences between patients and older control participants.

3.5. DDM Parameters Are Predicted by Clinical Assessments of Visual Perception

Data from both possible and probable MCI-LB patients were entered into a series of hierarchical stepwise linear regressions. First, demographic information including education, gender, age, and MCI-LB group (probable, possible) were entered as predictors of either ACE subscales, NEVHI, CAF, or DCFS scores. Following this, DDM parameters were added to the models as predictors.

A diagnosis of probable MCI-LB (adj R^2 = 0.12, beta = −0.30, p = 0.035) along with years of education (adj R^2 = 0.05, beta = 0.31, p = 0.028) were significant predictors for the ACE visuospatial score. The addition of DDM parameters to the model revealed that non-decision time (F (3, 45) = 5.413, R^2 = 0.265, adj R^2 = 0.129, beta = −0.333, p = 0.014) significantly predicted ACE visuospatial performance in the patient groups. The addition of non-decision time as a predictor in the model was also significant (Delta R^2 = 0.148, F change (1, 47) = 8.136, p = 0.006).

Moreover, male gender (adj R^2 = 0.06, beta = −0.295, =0.036) to significantly predicted NEVHI score. The addition of DDM parameters to the model showed that non-decision time approached significance but did not survive corrections for multiple comparisons (adj R^2 = 0.062, beta = 0.249, p = 0.06). Non-decision time was returned as a predictor of DCFS score but did not reach statistical significance (p = 0.072). Moreover, drift rate was returned as a predictor of CAF score, but did not reach statistical significance (p = 0.321).

Spearman's Rho correlational analysis showed that non-decision time was negatively correlated with ACE Visuo-spatial subscale (rho = −0.38, $p < 0.001$), but drift rate (p = 0.55) and boundary separation value (p = 0.92) were not (Figure 4). Mean non-decision time was associated with NEVHI score but this was not significant after correction for multiple comparisons (p = 0.06). Drift rate (p = 0.32) and boundary separation value (p = 0.45) did not show a significant relationship with NEVHI score. Importantly, UPDRS total motor impairment score did not significantly correlate with non-decision time (p = 0.23). This indicates that non-decision time is more strongly related to perceptual measures, rather than measures of motor impairments in this instance.

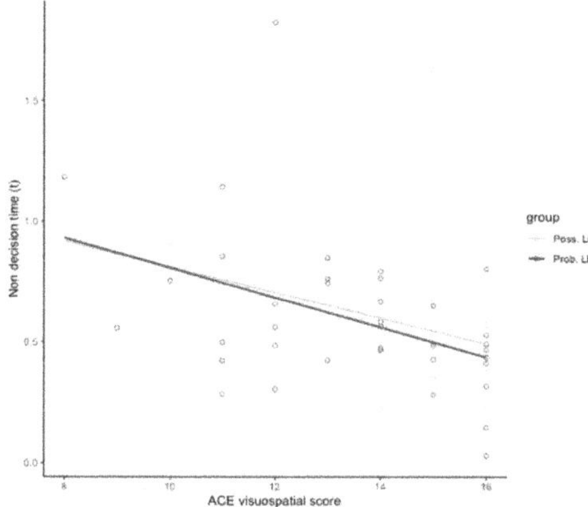

Figure 4. Non-decision time (t) is negatively related to scores on the Addenbrooke's Cognitive Exam (ACE) Visuospatial subscale in mild cognitive impairment (MCI-LB) groups (adj R^2 = 0.129, beta = −0.333, p = 0.014).

3.6. Clinical Measure of Visual Hallucinations Is Related to Non-Decision Time

To further investigate the relationship between visual hallucinations (VH) and drift diffusion parameters, group comparisons between those patients with a higher NEVHI score (NEVHI+), and those with a low NEVHI score (NEVHI−) were carried out. There was a significant difference in non-decision time between the NEVHI+ and NEVHI− group (U = 183, p = 0.024). Boundary separation (p = 0.416) and drift rate (p = 0.476) did not differ between groups. Finally, Spearman's Rho correlation showed a significant positive relationship between NEVHI score and non-decision time in MCI-LB patients in the NEVHI+ (rho = 5.412, p = 0.013) (Figure 5). However, as the NEVHI score is not a direct indicator of the presence of visual hallucinations, we also carried out a comparison between those patients with the presence of complex VH (VH+), as rated by a clinical panel—as per Donaghy et al. [24]—in accordance with 2017 DLB criteria [41], and patients with absence of complex VH (VH−). VH+ and VH− patients did not show any significant difference in non-decision time (p = 0.625) or drift rate (p = 0.659), but VH+ patients had significantly higher boundary separation values (U = 90, p = 0.002).

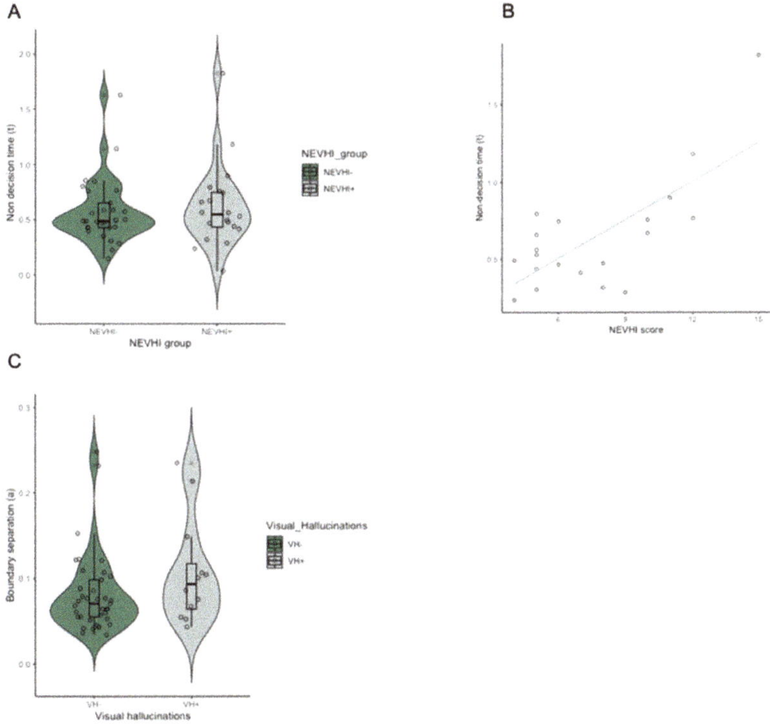

Figure 5. (**A**) Non-decision time (t) comparisons between NEVHI+ and NEVHI− MCI-LB patients (U = 183, p = 0.024). (**B**) Relationship between visual hallucinations score (NEVHI) and non-decision time in VH group (rho = 5.412, p = 0.013). (**C**) Boundary separation (a) comparisons between VH+ and VH− MCI-LB patients (U = 90, p = 0.002).

4. Discussion

Our results showed that mean RT were longer in the Lewy body groups than in the controls, however RT did not differ between possible and probable Lewy body groups, though group numbers were modest. Moreover, accuracy was lower in both Lewy body groups in comparison to controls.

These findings are consistent with previous reports of lengthened RTs and reduced accuracy in CRT for Lewy body patients [41,42]. Moreover, previously reported DDM profiles of healthy older adults were characterized by higher boundary separation values with preserved drift rate and non-decision times relative to younger adults. Our findings are consistent with previous findings in that drift rate, and therefore information processing ability, was unaffected by age, while older adults appear to wait to collate more information about the stimuli to ensure an accurate response [17]. This pattern reflects a more cautious strategy to decision making.

In addition, we found that MCI-LB patients had longer non-decision times and lower boundary separation values than healthy older controls but did not differ with regards to drift rate. We propose that this pattern of results suggests that patients had difficulties at perceptual encoding (non-decision) stages of decision making. This suggests that, rather than having impaired overall processing, Lewy body patients have more specific processing deficits in non-decision time, leading to longer RT. Moreover, patients also experienced shorter boundary separation values, suggesting that they adopted a less cautious approach than healthy older adults in their decision making in response to impaired perceptual encoding. This in turn may contribute to lower accuracy. However, no differences were observed in mean drift rates, although it is worthwhile mentioning that patients as a group exhibited increased variability in all DDM parameters. Thus, despite the null effect for mean drift rates, MCI-LB patients drift rate values were more heterogenous, suggesting that perceptual encoding and shorter boundary separation values also affected the rate of information accumulation and hence the quality of the decision-making process. Previous findings in hallucinating PD patients have reported lower drift rates, which indicated less effective sensory accumulation of evidence [19]. Here we did not find any differences in mean but in the variance of the drift rate which suggests that MCI-LB was associated with a greater variability of evidence accumulation. This may mean that MCI-LB patients experienced fluctuations in their ability to effectively process information, in part as a result of perceptual encoding problems, or in part as a result of impairments in alerting and focusing attention. Greater trial-to-trial variability in drift rates would result in slower RT and reduced accuracy, therefore investigating this further may be valuable in future research. As we employed a simple drift diffusion model due to limited trial numbers, this is difficult to recover, therefore future investigations should employ more complex models to examine the relationship between trial-by-trial variability in drift rates and fluctuation in attention. In addition, correlations between DDM variance and cognitive fluctuations may also be of interest, given the transient and variable nature of cognition in MCI-LB and DLB. Moreover, this may further illustrate the variability of cognitive impairment, and its relationship with subsequent clinical presentation in mild cognitive impairment stages of DLB [7].

Approximately 80% of DLB patients experience visual hallucinations [23] that are known to be related to impairments in visuospatial performance (Hamilton et al., 2012). Consistent with a previous study in PD [19] we observed differences between MCI-LB and control groups in non-decision time, a parameter that reflects both perceptual encoding and motor aspects. However, in contrast to the MCI-LB patients in the present study, PD patients had shorter, not longer, non-decision times in comparison to older adults [19]. As non-decision time is comprised of both perceptual and motor aspects, the authors proposed that shorter non-decision times may be partly due to impulsivity-related reductions in response latencies in PD. In the present study, non-decision times were not related to motor impairments, as measured by UPDRS score, but did relate to ACE visuospatial scores, suggesting that differences in MCI-LB patients were most likely due to impaired perceptual encoding. Together, these results suggest that, while the precise mechanisms may differ, both PD and MCI-LB patients experience perceptual encoding impairment that make sensory evidence less informative to them.

The results from our hierarchical stepwise regression fitting revealed that mean non-decision values predicted visuo-spatial performance but not NEVHI score. However, further analyses showed that patients in the NEVHI+ had longer non-decision times than NEVHI− and non-decision time correlated positively with higher NEVHI score in this group. As a positive NEVHI score does not necessarily reflect the incidence of visual hallucinations but can be due to other visual problems [43], we also classified patients on the basis of clinical ratings into those with a history of visual hallucinations and those without. This comparison revealed that individuals with visual hallucinations showed wider boundary separation than those without. Our finding of significantly longer non-decision times in the NEVHI+ versus the NEVHI− group suggests that patients who experience visual difficulties including visual hallucinations are prone to difficulties in perceptual encoding and that these encoding problems may contribute to their experience of visual hallucinations. The additional finding that individuals with a history of visual hallucinations showed larger boundary separation than those without, suggests that these individuals experience additional decision-making problems and adopt a more cautious response strategy in light of noisy perceptual input, that may contribute to and/or reflect the experience of visual hallucinations. Thus, perceptual encoding difficulties appear to be a necessary but not a sufficient condition for the occurrence of visual hallucinations. Such an interpretation is consistent with current theories of visual hallucinations in DLB [21,22] proposing that hallucinations occur as a product of impaired perceptual encoding or bottom-up processing, and subsequent over activation of top down processes, which may result in incorrect objects being inserted into the visual scene, i.e., visual hallucinations. In this context, our results may demonstrate that the processing of sensory stimuli is slower and potentially less informative in all MCI-LB patients but that in patients who hallucinate such impaired sensory processing is accompanied by additional top-down decision-making problems.

In contrast to our findings, O'Callaghan et al. [19] reported that all PD patients showed wider decision boundaries, but only hallucinating patients showed slower drift rates than those patients who did not hallucinate. While this pattern of results also supports the view that impaired sensory evidence accumulation may lead to reduced information processing quality and an over-reliance on top down processing in PD, which in turn may underpin visual hallucinations, the mechanisms leading to these impairments seem to differ between PD and MCI-LB.

There were limitations of the present study. Due to practicalities in recruitment, healthy control participants were significantly younger on average than probable and possible Lewy body patients, which may have affected some drift diffusion estimates. However, age was not a significant predictor in the linear regression analyses. In addition, due to recruitment from a University panel, control participants had significantly more years of education than Lewy body groups, which may also have influenced performance in some cognitive tests. We assessed the potential influence of age and education on our results by comparing patients with a subgroup of controls that were closely matched for these variables. These additional analyses resulted in the same pattern of results, suggesting that the group differences in age and education did not drive our findings.

Moreover, although tasks were similar in nature between healthy control and patient groups, there were some notable differences in both trial number and stimuli which could influence RT and accuracy and bias the DDM modelling. Firstly, trial numbers were relatively low in the CRT in MCI-LB groups, due to practical limitations and patients' task completion ability, which may influence the accuracy of model fitting. Secondly, there is possibility that the presence of neutral flankers in the healthy control task may produce lower drift rates due to crowding effects, therefore resulting in an under estimation of the current group differences in our results. Thirdly, as control participants also took part in randomly presented congruent and incongruent trials, boundary separation may be raised as older adults become more cautious in response overall, thus overestimating the boundary separation effect in this study [44]. Finally, non-decision time may be influenced by a more complex stimuli, resulting in an under-estimation of the observed perceptual encoding group differences in the present study. This does pose some limitation to the conclusions which can be drawn, therefore replication in a larger sample, and with identical task conditions is vital to clarify these effects. Despite this, employing the same task between patients and control subjects is also challenging, as those tasks which are accessible for patients in terms of trial number or stimuli may be very simple for controls to complete, resulting in a ceiling effect, and vice versa.

To conclude, the present findings suggest that impaired perceptual encoding, as estimated by the DDM, contributes to NEVHI score in MCI-LB, and altered boundary separation may be related to the presence of visual hallucinations [19,45].

Author Contributions: Conceptualization, L.R. and C.M.-B.; Data curation, P.C.D.; Formal analysis, L.R.; Resources, C.A.H., P.C.D., and A.T.; Supervision, C.M.-B.; Writing—original draft, L.R.; Writing—review and editing, C.A.H., J.C., P.C.D., A.T., and C.M.-B. All authors have read and agreed to the published version of the manuscript.

Funding: LR is supported by a School of Psychology Open PhD scholarship. This research was supported by the NIHR Newcastle Biomedical Research Centre (grant numbers BH120812 and BH120878).

Acknowledgments: We gratefully acknowledge colleagues at Newcastle University involved in the LewyPro study for provision of patient data. The authors would also like to acknowledge Craig Hedge for his generous provision of EZ scripts, and his insightful help and comments.

Conflicts of Interest: The authors declare no conflict of interest.

References

1. McKeith, I.G.; Boeve, B.F.; Dickson, D.W.; Halliday, G.; Taylor, J.-P.; Weintraub, D.; Aarsland, D.; Galvin, J.; Attems, J.; Ballard, C.G.; et al. Diagnosis and management of dementia with Lewy bodies: Fourth consensus report of the DLB Consortium. *Neurology* **2017**, *89*, 88–100. [CrossRef]
2. Metzler-Baddeley, C. A Review of Cognitive Impairments in Dementia with Lewy Bodies Relative to Alzheimer's Disease and Parkinson's Disease with Dementia. *Cortex* **2007**, *43*, 583–600. [CrossRef]
3. Ciafone, J.; Little, B.; Thomas, A.J.; Gallagher, P. The Neuropsychological Profile of Mild Cognitive Impairment in Lewy Body Dementias. *J. Int. Neuropsychol. Soc.* **2020**, *26*, 210–225. [CrossRef] [PubMed]
4. Collerton, D.; Burn, D.; McKeith, I.; O'Brien, J. Systematic Review and Meta-Analysis Show that Dementia with Lewy Bodies Is a Visual-Perceptual and Attentional-Executive Dementia. *Dement. Geriatr. Cogn. Disord.* **2003**, *16*, 229–237. [CrossRef] [PubMed]
5. Bowman, A.; Bruce, V.; Colbourn, C.J.; Collerton, D. Compensatory shifts in visual perception are associated with hallucinations in Lewy body disorders. *Cogn. Res. Princ. Implic.* **2017**, *2*, 26. [CrossRef] [PubMed]
6. Taylor, J.-P.; Colloby, S.J.; McKeith, I.G.; O'Brien, J.T. Covariant perfusion patterns provide clues to the origin of cognitive fluctuations and attentional dysfunction in Dementia with Lewy bodies. *Int. Psychogeriatr.* **2013**, *25*, 1917–1928. [CrossRef]
7. McKeith, I.G.; Ferman, T.J.; Thomas, A.J.; Blanc, F.; Boeve, B.F.; Fujishiro, H.; Kantarci, K.; Muscio, C.; O'Brien, J.; Postuma, R.B.; et al. Research criteria for the diagnosis of prodromal dementia with Lewy bodies. *Neurology* **2020**, *94*, 743–755. [CrossRef] [PubMed]
8. Gauthier, S.; Reisberg, B.; Zaudig, M.; Petersen, R.C.; Ritchie, K.; Broich, K.; Belleville, S.; Brodaty, H.; Bennett, D.; Chertkow, H.; et al. Mild cognitive impairment. *Lancet* **2006**, *367*, 1262–1270. [CrossRef]
9. Ballard, C.; O'Brien, J.; Morris, C.M.; Barber, R.; Swann, A.; Neill, D.; McKeith, I. The progression of cognitive impairment in dementia with Lewy bodies, vascular dementia and Alzheimer's disease. *Int. J. Geriatr. Psychiatry* **2001**, *16*, 499–503. [CrossRef]
10. Bradshaw, J. Fluctuating cognition in dementia with Lewy bodies and Alzheimer's disease is qualitatively distinct. *J. Neurol. Neurosurg. Psychiatry* **2004**, *75*, 382–387. [CrossRef]
11. Fuentes, L.J.; Fernández, P.J.; Campoy, G.; Antequera, M.M.; García-Sevilla, J.; Antúnez, C. Attention Network Functioning in Patients with Dementia with Lewy Bodies and Alzheimer's Disease. *Dement. Geriatr. Cogn. Disord.* **2010**, *29*, 139–145. [CrossRef] [PubMed]
12. Bailon, O.; Roussel, M.; Boucart, M.; Krystkowiak, P.; Godefroy, O. Psychomotor Slowing in Mild Cognitive Impairment, Alzheimer's Disease and Lewy Body Dementia: Mechanisms and Diagnostic Value. *Dement. Geriatr. Cogn. Disord.* **2010**, *29*, 388–396. [CrossRef] [PubMed]
13. Cormack, F.; Gray, A.; Ballard, C.; Tovée, M.J. A failure of'Pop-Out' in visual search tasks in dementia with Lewy Bodies as compared to Alzheimer's and Parkinson's disease. *Int. J. Geriatr. Psychiatry* **2004**, *19*, 763–772. [CrossRef] [PubMed]
14. Ratcliff, R. A theory of memory retrieval. *Psychol. Rev.* **1978**, *85*, 59–108. [CrossRef]
15. Ratcliff, R.; Rouder, J.N. Modeling Response Times for Two-Choice Decisions. *Psychol. Sci.* **1998**, *9*, 347–356. [CrossRef]

16. Wiecki, T.V.; Sofer, I.; Frank, M.J. HDDM: Hierarchical Bayesian estimation of the Drift-Diffusion Model in Python. *Front. Neuroinform.* **2013**, *7*, 14. [CrossRef]
17. Ratcliff, R. Modeling response signal and response time data. *Cogn. Psychol.* **2006**, *53*, 195–237. [CrossRef]
18. Spaniol, J.; Madden, D.J.; Voss, A. A Diffusion Model Analysis of Adult Age Differences in Episodic and Semantic Long-Term Memory Retrieval. *J. Exp. Psychol. Learn. Mem. Cogn.* **2006**, *32*, 101–117. [CrossRef]
19. O'Callaghan, C.; Hall, J.M.; Tomassini, A.; Muller, A.J.; Walpola, I.C.; Moustafa, A.A.; Shine, J.M.; Simon, J.G.L. Visual Hallucinations Are Characterized by Impaired Sensory Evidence Accumulation: Insights from Hierarchical Drift Diffusion Modeling in Parkinson's Disease. *Biol. Psychiatry Cogn. Neurosci. Neuroimaging* **2017**, *2*, 680–688.
20. Wagenmakers, E.-J.; Van Der Maas, H.L.J.; Grasman, R.P.P.P. An EZ-diffusion model for response time and accuracy. *Psychon. Bull. Rev.* **2007**, *14*, 3–22. [CrossRef]
21. Collerton, D.; Perry, E.; McKeith, I. Why people see things that are not there: A novel Perception and Attention Deficit model for recurrent complex visual hallucinations. *Behav. Brain Sci.* **2005**, *28*, 737–757. [CrossRef] [PubMed]
22. Shine, J.M.; Halliday, G.M.; Naismith, S.L.; Lewis, S.J.G. Visual misperceptions and hallucinations in Parkinson's disease: Dysfunction of attentional control networks? *Mov. Disord.* **2011**, *26*, 2154–2159. [CrossRef] [PubMed]
23. Onofrj, M.; Taylor, J.P.; Monaco, D.; Franciotti, R.; Anzellotti, F.; Bonanni, L.; Onofrj, V.; Thomas, A. Visual Hallucinations in PD and Lewy Body Dementias: Old and New Hypotheses. *Behav. Neurol.* **2013**, *27*, 479–493. [CrossRef] [PubMed]
24. Donaghy, P.C.; Taylor, J.-P.; O'Brien, J.; Barnett, N.; Olsen, K.; Colloby, S.J.; Lloyd, J.; Petrides, G.; McKeith, L.G.; Thomas, A.J. Neuropsychiatric symptoms and cognitive profile in mild cognitive impairment with Lewy bodies. *Psychol. Med.* **2018**, *48*, 2384–2390. [CrossRef] [PubMed]
25. Thomas, A.J.; Donaghy, P.C.; Roberts, G.; Colloby, S.J.; Barnett, N.A.; Petrides, G.; Lloyd, J.; Olsen, K.; Taylor, J.-P.; McKeith, L.G.; et al. Diagnostic accuracy of dopaminergic imaging in prodromal dementia with Lewy bodies. *Psychol. Med.* **2019**, *49*, 396–402. [CrossRef] [PubMed]
26. Mioshi, E.; Dawson, K.; Mitchell, J.; Arnold, R.; Hodges, J.R. The Addenbrooke's Cognitive Examination Revised (ACE-R): A brief cognitive test battery for dementia screening. *Int. J. Geriatr. Psychiatry* **2006**, *21*, 1078–1085. [CrossRef]
27. Warrington, E.K.; McKenna, P. *Graded Naming Test: Manual*; Cambridge Cognition Ltd.: Bottisham, UK, 1983.
28. Rey, A. *[The Clinical Examination in Psychology]*; Presses Universitaries De France: Oxford, UK, 1958; 222p. (L'examen Clinique en Psychologie).
29. Yesavage, J.A. Geriatric Depression Scale. *Psychopharmacol. Bull.* **1988**, *24*, 709–711.
30. Walker, M.P.; Ayre, G.; Perry, E.; Wesnes, K.A.; McKeith, L.G.; Tovée, M.J.; Edwardson, J.; Ballard, C. Quantification and Characterisation of Fluctuating Cognition in Dementia with Lewy Bodies and Alzheimer's Disease. *Dement. Geriatr. Cogn. Disord.* **2000**, *11*, 327–335. [CrossRef]
31. Lee, D.R.; McKeith, I.G.; Mosimann, U.P.; Ghosh-Nodial, A.; Grayson, L.; Wilson, B.; Thomas, A.J. The Dementia Cognitive Fluctuation Scale, a New Psychometric Test for Clinicians to Identify Cognitive Fluctuations in People with Dementia. *Am. J. Geriatr. Psychiatry* **2014**, *22*, 926–935. [CrossRef]
32. Cummings, J.L.; Mega, M.; Gray, K.; Rosenberg-Thompson, S.; Carusi, D.A.; Gornbein, J. The Neuropsychiatric Inventory: Comprehensive assessment of psychopathology in dementia. *Neurology* **1994**, *44*, 2308. [CrossRef]
33. Mosimann, U.P.; Collerton, D.; Dudley, R.; Meyer, T.D.; Graham, G.; Dean, J.L.; Bearn, D.; Killen, A.; Dickinson, L.; Clarke, M.P.; et al. A semi-structured interview to assess visual hallucinations in older people. *Int. J. Geriatr. Psychiatry* **2008**, *23*, 712–718. [CrossRef] [PubMed]
34. Goetz, C.G.; Fahn, S.; Martínez-Martín, P.; Poewe, W.; Sampaio, C.; Stebbins, G.T.; Stern, M.B.; Tilley, B.C.; Dodel, R.; Dubois, B.; et al. Movement Disorder Society-sponsored revision of the Unified Parkinson's Disease Rating Scale (MDS-UPDRS): Process, format, and clinimetric testing plan. *Mov. Disord.* **2007**, *22*, 41–47. [CrossRef] [PubMed]
35. Lawton, M.P.; Brody, E.M. Assessment of Older People: Self-Maintaining and Instrumental Activities of Daily Living. *Gerontologist* **1969**, *9*, 179–186. [CrossRef] [PubMed]
36. Eriksen, B.A.; Eriksen, C.W. Effects of noise letters upon the identification of a target letter in a nonsearch task. *Percept. Psychophys.* **1974**, *16*, 143–149. [CrossRef]

37. van Ravenzwaaij, D.; Oberauer, K. How to use the diffusion model: Parameter recovery of three methods: EZ, fast-dm, and DMAT. *J. Math. Psychol.* **2009**, *53*, 463–473. [CrossRef]
38. Ratcliff, R.; McKoon, G. The Diffusion Decision Model: Theory and Data for Two-Choice Decision Tasks. *Neural Comput.* **2008**, *20*, 873–922. [CrossRef] [PubMed]
39. Ratcliff, R.; Childers, R. Individual Differences and Fitting Methods for the Two-Choice Diffusion Model of Decision Making. *Decision (Wash.)*. 2015. Available online: https://www.ncbi.nlm.nih.gov/pmc/articles/PMC4517692/ (accessed on 13 May 2020).
40. Benjamini, Y.; Hochberg, Y. Controlling the False Discovery Rate: A Practical and Powerful Approach to Multiple Testing. *J. R. Stat. Soc. Ser. B (Methodol.)* **1995**, *57*, 289–300. [CrossRef]
41. Firbank, M.; Kobeleva, X.; Cherry, G.; Killen, A.; Gallagher, P.; Burn, D.J.; Thomas, A.J.; O'Brien, J.; Taylor, J.-P. Neural correlates of attention-executive dysfunction in lewy body dementia and Alzheimer's disease: Neural Correlates of Attention in Dementia. *Hum. Brain Mapp.* **2015**, *37*, 1254–1270. [CrossRef]
42. Firbank, M.J.; Parikh, J.; Murphy, N.; Killen, A.; Allan, C.L.; Collerton, D.; Blamire, A.; Taylor, J.-P. Reduced occipital GABA in Parkinson disease with visual hallucinations. *Neurology* **2018**, *91*, e675–e685. [CrossRef]
43. Holiday, K.A.; Pirogovsky-Turk, E.; Malcarne, V.L.; Filoteo, J.V.; Litvan, I.; Lessig, S.L.; Song, D.; Schiehser, D.M. Psychometric Properties and Characteristics of the North-East Visual Hallucinations Interview in Parkinson's Disease. *Mov. Disord. Clin. Pract.* **2017**, *4*, 717–723. [CrossRef]
44. Hedge, C.; Powell, G.; Sumner, P. The mapping between transformed reaction time costs and models of processing in aging and cognition. *Psychol. Aging* **2018**, *33*, 1093. [CrossRef] [PubMed]
45. Diederich, N.J.; Goetz, C.G.; Stebbins, G.T. Repeated visual hallucinations in Parkinson's disease as disturbed external/internal perceptions: Focused review and a new integrative model. *Mov. Disord.* **2005**, *20*, 130–140. [CrossRef] [PubMed]

Article

The Relation between Sustained Attention and Incidental and Intentional Object-Location Memory

Efrat Barel * and Orna Tzischinsky

Department of Behavioral Sciences, The Max Stern Academic College of Emek Yezreel, Emek Yezreel 19300, Israel; orna@yvc.ac.il

* Correspondence: efratb@yvc.ac.il

Received: 29 January 2020; Accepted: 29 February 2020; Published: 4 March 2020

Abstract: The role of attention allocation in object-location memory has been widely studied through incidental and intentional encoding conditions. However, the relation between sustained attention and memory encoding processes has scarcely been studied. The present study aimed to investigate performance differences across incidental and intentional encoding conditions using a divided attention paradigm. Furthermore, the study aimed to examine the relation between sustained attention and incidental and intentional object-location memory performance. Based on previous findings, an all women sample was recruited in order to best illuminate the potential effects of interest. Forty-nine women participated in the study and completed the psychomotor vigilance test, as well as object-location memory tests, under both incidental and intentional encoding divided attention conditions. Performance was higher in the incidental encoding condition than in the intentional encoding condition. Furthermore, sustained attention correlated with incidental, but not with intentional memory performance. These findings are discussed in light of the automaticity hypothesis, specifically as it regards the role of attention allocation in encoding object-location memory. Furthermore, the role of sustained attention in incidental memory performance is discussed in light of previous animal and human studies that have examined the brain regions involved in these cognitive processes. We conclude that under conditions of increased mental demand, executive attention is associated with incidental, but not with intentional encoding, thus identifying the exact conditions under which executive attention influence memory performance.

Keywords: object-location memory; sustained attention; incidental encoding; intentional encoding

1. Introduction

Object-location memory is a complex neurocognitive ability that presents a challenge for our cognitive system. It involves three components: (1) object-processing, (2) spatial-location processing, and (3) object-location binding [1]. Object-location memory is a fundamental ability that is needed in our daily lives. Given its adaptive value for both humans and animals, it has been suggested that object-location memory is not only driven by conscious recollections of objects' locations, but rather that it is an automatic process and possibly influenced by unconscious memory [2].

The automaticity hypothesis [2], in regard to the role of attention allocation in encoding object-location memory, has been widely studied by comparing memory performance across *incidental* and *intentional* encoding conditions. In intentional encoding conditions, participants are explicitly instructed about a required subsequent retrieval phase, whereas in incidental encoding conditions, participants are shown an array of stimuli without awareness of a subsequent retrieval phase. Findings regarding memory performance are inconsistent; some studies have shown that the locations of objects are learned without participants receiving explicit instruction to remember the locations (e.g., [3]), thus supporting the automaticity hypothesis. Other studies have shown that intention to remember locations improves memory performance (e.g., [4]). Postma and colleagues [1] speculated that each component of object-location memory differs in its processing automaticity, with the spatial-location component operating more automatically than the object identity and object-location binding processing components.

Other attention allocation tasks involve differentiating between *divided* versus *full* attention. In divided attention tasks, participants are required to respond to both target and distractor stimuli, whereas in full attention tasks, participants are required to direct their attention to the target stimulus only [5]. Studies investigating memory have typically demonstrated that in various memory tasks, divided attention during incidental or intentional encoding reduces performance (e.g., [5–8]). However, more recently, it has been suggested that under specific conditions, divided attention may facilitate memory performance. For example, Nussenbaum, Amso, and Markant [9] have shown that increasing the number of distractors in a divided attention condition did not impair memory for the target content. Furthermore, when distractors contained information that conflicted with the target content, increasing the number of distractors actually enhanced participants' memory.

The effects of attention during object-location memory encoding have been studied in the realm of visual working memory. Visual working memory is mainly characterized by its limited capacity, therefore the maintenance of attention to visual items is important in our daily behavior [10]. Previous studies investigated whether features and locations are represented as integrated objects in our visual working memory under various attention conditions. For example, Treisman and Zhang [11] demonstrated that attended objects are bound to their locations, however visual memory for binding is not disrupted when attention is directed to irrelevant stimuli. As opposed to working memory, long term memory requires different cognitive mechanisms at encoding, storage, and retrieval. The role of attention during object-location long-term memory encoding has been scarcely studied. To our knowledge, only two studies have examined the role of attention during object-location memory encoding under divided and full attention conditions. One study presented participants with an array of actual objects, and incorporated a verbal arithmetic task as a distraction in the divided attention condition [12]. The second study used a paper-and-pencil task and a tone discrimination task as an auditory distraction [13]. Both studies demonstrated that, with intentional encoding instruction, participants performed better in the full attention condition than the divided attention condition. However, under the incidental encoding conditions, this finding was replicated only in the study that used the actual array of objects, and not in the paper-and-pencil paradigm study. Furthermore, in the paper-and-pencil study women performed better in incidental than in intentional memory. Ecuyer-Dab and Robert [14] suggest that women might employ different strategies for incidental and intentional encoding. While under incidental encoding conditions, women spontaneously encode the surrounding elements, under intentional encoding with explicit instructions to memorize objects, women use an alternative strategy. In other words, females and males differ in the attentional and perceptual mechanisms that they employ in memorizing objects locations. Under incidental conditions, females unintentionally encode detailed features of their surroundings, and later are able to retrieve a precise representation of it. In contrast, under intentional conditions females employ a different strategy, such as verbal labeling of stimuli. For example, Lewin and colleagues [15] failed to demonstrate sex differences in location memory for uncommon objects and speculated that the absence of the familiar female superiority was due to females' difficulty in assigning verbal labels to the objects. In accordance with previous studies examining the role of attention in various memory tasks, these findings suggest that the influence of attention allocation on memory performance is not uniform, but rather that it is affected by the nature of the distractors—including the modality and the relatedness of the distractor to the target—as well as the nature and the modality of the target.

Studies investigating the neural correlates of object-location memory have demonstrated the involvement of the right hippocampus in spatial binding [1]. Only a few studies have examined the neural correlates of implicit and explicit spatial memory. Whereas some studies have shown the importance of the medial temporal lobes [16] and the striatum [17] for implicit spatial memory, others have demonstrated the importance of the diencephalic and frontal regions for explicit object-location memory [18].

The role of attention in memory performance has been also studied through a central component of executive function: *executive attention* or *attention control*. Attention control refers to attentional processes that support the ability to sustain information in the presence of internal or external distractions [16]. There are several attention control abilities, including, attention restraint, attention constraint [17], and sustained attention [16]. Sustained attention refers to attention control processes needed to preserve attention and task engagement over time (also referred to as vigilant attention; [18–20]). Studies that have addressed the relation between sustained attention, measured by the psychomotor vigilance task, and working memory capacity revealed that sustained attention was positively correlated with working memory capacity (e.g., [21]). Unsworth and Robison [16] recently proposed a cognitive-energetic model to explain the relation between sustained attention and various cognitive constructs, including memory. The underlying notion of the model is that intensity of attention varies within and between individuals. The intensity of attention is influenced by both intrinsic and extrinsic motivation levels, overall arousal levels, and intrinsic alertness. When attention intensity levels are high, task engagement is high and control levels are optimal. In four experiments examining the relation between sustained attention and working memory capacity, Unsworth and Robison [16] showed that this relationship is mediated by variation in intrinsic alertness—the ability to voluntarily control the intensity of attention on a continuous basis.

In the search for the underlying cause of reduced sustained attention, a phenomenon called vigilance decrement, two theories have been suggested: the under-load theory and the over-load theory [22–24]. In the under-load theory, the decrement is deemed to be due to boredom, mindlessness, or goal habituation [25], whereas in the over-load theory, vigilance decrement is considered to be due to mental fatigue and resource depletion. In order to examine these two theories, a few studies sought to investigate the influence of various working memory demands on vigilance decrement. For example, in Helton and Russell's [26] study, participants performed a vigilance task while simultaneously performing either a verbal or spatial working memory task. The researchers found that the concurrent verbal and spatial working memory load impacted the vigilance decrement among the participants. They concluded that vigilance decrement was caused by high cognitive resource demands, thus supporting the over-load theory.

The role of executive attention has also been examined in relation to incidental and intentional memory. Kontaxopoulou and colleagues [27] assessed participants' episodic memory performance via virtual reality stimuli, both incidentally and intentionally, using both verbal and visuospatial tests. Additionally, participants completed a neuropsychological battery assessing attention and executive functioning. The researchers found that almost all attentional and executive functioning measures were associated with participants' incidental, but not intentional, memory performance. They further reported that aging affected incidental (but not intentional) encoding processes. Given these two findings, Kontaxopoulou and colleagues [27] proposed that the ability to effectively execute incidental memory processes is more strongly connected with the overall cognitive system than is the ability to carry out intentional memory processes, as indicated by the association found with attention and executive functions. Indeed, memory studies among aging populations have shown that low scores on memory tasks were correlated with reduced activation in the frontal lobes (e.g., [28]). Furthermore, imaging studies have shown that there is a positive correlation between executive functioning and prefrontal cortex volume (for a meta-analysis, [29]). Therefore, it is suggested that incidental encoding processes, which hold a more prominent function in our daily lives, are perhaps more influenced by executive attention than intentional encoding processes

The role of attentional resources in memory performance has been previously investigated, especially among aging populations. The search for the source of memory decline in some memory functions, but not in others, led researchers to examine several hypotheses to explain the underlying mechanisms in memory decline [30]. One of the hypotheses concerns the role of reduction in attentional resources in episodic memory (e.g., [30,31]), and the emphasis on stability of attention during encoding of items in recall and recognition tasks [32]. Previous studies demonstrated that aging is associated with long-term episodic memory decline, including object-location memory, and is characterized by reduction in attentional resources in episodic memory. Therefore, the present study aims at fine-tuning the conditions under which attentional resources influence memory performance among young adults, in order to uncover the processing mechanisms needed for executing essential functions in our daily lives, such as object-location memory. Specifically, the present study aimed to explore the role that executive attention, especially sustained attention, plays in incidental and intentional object-location memory performance. In line with the over-load theory, the present study was conducted under conditions of divided attention. Increasing mental demand among the participants enabled us to pinpoint the exact conditions under which vigilance decrement influenced performance on incidental and intentional memory encoding tasks. To the best of our knowledge, this study is the first to examine the association between these variables. Whereas a previous study [27] investigated the role of executive attention in relation to incidental and intentional memory (verbal and visuospatial), the present study is the first to examine this relation with the outcome of object-location memory, and under conditions of cognitive load.

Furthermore, previous studies have demonstrated that getting a sufficient amount of sleep per night, on a regular basis, has an important influence on behavioral alertness and cognitive performance [33]. Therefore, in the current study, the quality and quantity of sleep during the four nights prior to the experiment were measured and used as an indicator of sleep deprivation and fatigue on the morning of the experiment.

In summary, previous studies did not investigate object-location memory performance under cognitive load. Furthermore, to the best of our knowledge only one study assessed the role of attentional and executive functioning in incidental and intentional episodic memory [27]. However, it is not yet known whether sustained attention influences incidental and intentional episodic memory under cognitive load. Therefore, in the present study, including cognitive load allowed for an examination of whether sustained attention influences memory performance. Furthermore, the present study allowed us to examine whether sustained attention influences incidental as well as intentional encoding. Moreover, the present study controlled for the quality and quantity of sleep that have been related to cognitive performance.

Our main hypotheses are:

(1) Memory performance would be higher under incidental encoding as compared to intentional encoding.
(2) Sustained attention, as measured by the psychomotor vigilance test, would be associated with incidental, but not intentional, encoding measures.

2. Method

2.1. Participants

Forty-nine female students (mean age 24.5 ± 1.89) from a college in the north of Israel participated in the study. Participants were recruited through advertisements at the college, and received course credit for their participation. We chose to recruit a female sample given the extensive body of research suggesting females' superior performance in object-location memory tasks as compared to males (for a meta-analysis, see [34]).

2.2. Materials

The study was approved by the institutional review boards (IRBs) of the college (no: EMEK YVC2018-20). All participants arrived at the lab between 8:00 a.m. and 12:00 p.m., after four nights with ActiGraph, to participate in the experiment. After providing informed consent, participants completed a brief demographic questionnaire and a number of tasks. The tasks are outlined below.

Object location memory–Incidental encoding: The study included a stimulus array of 25 black-and-white drawings of objects, based on those used in the Eals and Silverman [35] study. The stimuli were presented on standard-size A4 white paper. Participants were instructed to complete both a pricing task and a distraction task within a one-minute time-frame. In the pricing task, which was designed to manipulate incidental (non-directed) encoding, participants were asked to write a price tag for each object directly on the paper. They were told that if they were unable to estimate a price for the object, they should provide a guess [36]. In the distraction task, designed to increase mental load, a pre-recorded soundtrack of piano tones, randomized by pitch (low or high), was presented in intervals of 2 or 3 s [37]. The participants were asked to indicate the low-pitched tone by raising their left hand and the high-pitched tone by raising their right hand. Immediately afterward, the participants were shown another stimulus array, in which 14 of the objects were in different locations than before. They were given 60 s to mark which objects' locations were unchanged and circle the ones whose positions had changed. A manipulation check indicated that the participants were not suspicious about the purpose of the experiment.

Object location memory–Intentional encoding: The design of the intentional encoding condition, including all of the materials presented, was identical to the incidental encoding task phase, with a new stimulus array. However, the one difference was that in the intentional encoding task, participants were given one minute to "try to memorize as many objects in the array as possible and their approximate locations" ([35], p. 100) before the distraction task was introduced.

In the present study, the paper-and-pencil format was chosen due to task requirements of the incidental encoding condition (participants were asked to write a price estimation per each item). Given the limited timeframe (60 s) and the potential variation in computer skills across participants, a paper-and-pencil format (previously used for the same study purposes, e.g., [37]) was utilized.

2.3. *Psychomotor Vigilance Test (PVT):*

Participants completed a visual psychomotor vigilance task (PVT). This task is sensitive to sleep loss and circadian phase [38,39]. The psychomotor vigilance task (PVT) has been employed for the last 30 years as a sensitive test of sustained attention [40]. This simple measure of reaction time (RT) to repetitive stimuli has become recognized as a highly sensitive effective tool for measuring degradation of sustained attention performance under sleep deprivation or change in circadian phase [38,39]. The PVT is the most widely used measure of behavioral alertness. The standard duration of the PVT is 10 min; however, the optimal duration of the PVT is shorter than 10 min. Most studies use PVT outcomes to monitor sensitivity to total and partial sleep loss [41,42], and to differentiate sleep-deprived subjects from alert subjects [43].

However, in the present study the use of the PVT was for a different purpose (besides evaluating sensitivity to sleep loss; [22]); the PVT was used to explore the role that sustained attention plays in incidental and intentional object location memory performance. The PVT-B is a validated measure of sustained attention, with high test–retest reliability and low learning effects [19]. Previous studies validated the PVT as sensitive in differentiating dementia patients from healthy controls, and has been used among aging and MCI populations [44,45].

The PVT-B (Joggle Research Program, Seattle, WA, USA) is a sustained attention reaction time task, and it was performed on an iPad in the current study. Participants were instructed to maintain *vigilant attention* on a target box and respond as quickly as possible to the appearance of a stimulus, while avoiding responding prematurely. The outcome measure of the current study was the *Aggregate Score.* A score metric that penalized participants based on the percentage of responses that were lapses and the percentage of responses that were considered to be early response errors was calculated. The calculation was as follows: Aggregate Score = (1 − (Lapses/Responses) − (Errors/Responses)) × 100.

With regard to executive attention, it is important to emphasize that there is extensive evidence that the neurobehavioral consequences of sleep loss can be measured through certain aspects of cognitive functioning [46–48]. Among the most reliable effects of sleep deprivation is degradation of attention [38,49], including vigilant attention [38,50].

For each trial, an empty box was presented on an iPad screen, triggering a millisecond counter. Participants were requested to press on the screen to stop the counter. Participants were instructed to respond as quickly as possible, but to avoid pressing on the screen when the counter was not displayed (i.e., false starts). The inter-stimulus interval, defined as the period between the last response and the appearance of the next stimulus, varied randomly from 2–10 s [49].

Sleep patterns–The purpose of the objective sleep test was to control the quality and quantity of sleep, and to ensure that subjects did not suffer from sleep deprivation during the four days prior to the study.

Objective sleep patterns were measured using an actigraph (AMI, NY). This small device measures sleep patterns in one's natural environment and provides objective data of one's sleep patterns. Participants wore the actigraph in the four nights preceding the experiment. Actigraph recordings provided an estimation of participants' sleep onset, wake time, sleep latency, sleep duration, true sleep minutes, wake after sleep onset (WASO), and sleep efficiency.

2.4. Procedure

Prior to arriving to the lab for the experiment, participants' objective sleep patterns were measured for four consecutive nights. Participants were then individually invited to the lab. All participants completed the study in the same room, The experiment utilized a crossover design, such that half of the participants performed the PVT-B task first, followed by the memory tasks, and the other half of the participants performed the tasks in the opposite order. Due to the study design (which included an incidental encoding condition), all participants performed the memory tasks (under incidental and intentional conditions) in the same order.

2.5. Power Analyses

G Power 3 (Heinrich Heine University, Düsseldorf, Germany) was used to determine the sample size required to find a significant difference in object-location memory performance between incidental and intentional encoding. The power analysis indicated that 47 participants would be needed to detect a medium effect size (dz = 0.5) with an alpha level of 0.05 and 90% power.

3. Results

3.1. Sleep Measures

Participants' objective sleep patterns were characterized by sleep measures that fell within the typical ranges, a sleep duration that matched the recommended duration, and a high quality of sleep (i.e., sleep efficiency was high; see Table 1).

Table 1. Actigraphic sleep pattern.

	Mean ± SD
Sleep Onset	00:22 ± 1.09
Wake Time	8:47 ± 0.89
Sleep Latency	13.92 ± 12.13
Sleep Duration	445.93 ± 65.74
True Sleep Minutes	410.35 ± 61.97
Waso (min)	15.83 ± 13.41
Sleep Efficiency (%)	96.17 ± 3.0

3.2. Incidental vs. Intentional Memory Performance

All the analyses were based on non-parametric statistics since location memory scores (incidental and intentional) were not normally distributed. Object-location memory performance across the incidental and intentional encoding conditions were calculated as two scores. The first was the total number of correct identifications of exchanged objects, as customarily used in the literature (e.g., [35,51]). The second was calculated as the number of correct identifications minus the number of incorrect identifications [52]. To test the difference in object-location memory performance across the incidental and intentional encoding conditions, we performed a Wilcoxon signed-rank test and found a significant effect, both for the total score ($Z = 4.72$, $p < 0.001$, *effect size* = 0.67) (see Figure 1), and for the corrected score ($Z = 3.82$, $p < 0.001$, *effect size* = 0.61) (see Figure 2). Participants scored higher on location memory in the incidental encoding condition as compared with the intentional encoding condition.

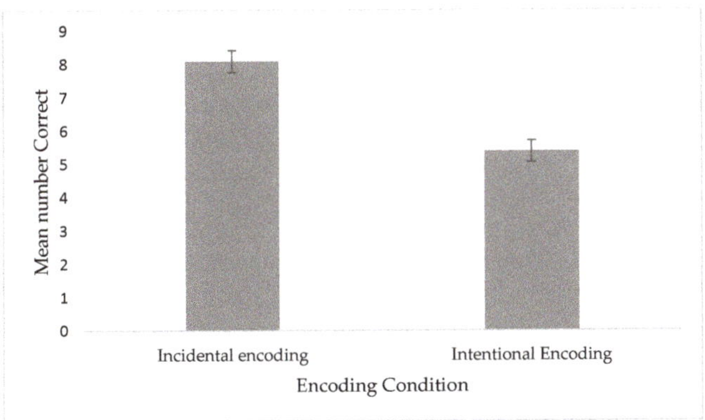

Figure 1. Mean number of correctly detected location-exchanged objects under divided attention incidental and intentional encoding conditions for the total scores. Error bars represent standard errors of the mean (SEM).

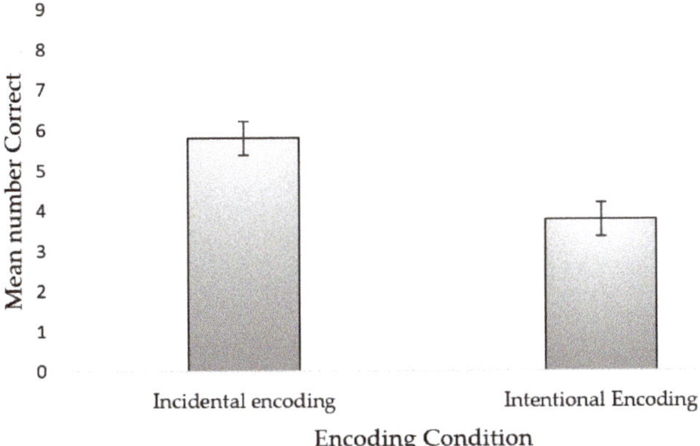

Figure 2. Mean number of correctly detected location-exchanged objects under divided attention incidental and intentional encoding conditions for the corrected scores. Error bars represent standard errors of the mean (SEM).

3.3. PVT-B Measures

Table 2 displays the mean and standard deviations of PVT-B measures.

Table 2. PVT-B measures.

	Mean ± SD
Responses	44.48 ± 2.21
Errors	2.37 ± 3.18
Aggregate Score	85.81 ± 13.77

3.4. The Role of Sustained Attention in Incidental vs. Intentional Memory Performance

First, correlation was performed between incidental and intentional memory performance. No significant correlation was found ($r = 0.27$, $p > 0.05$). Next, correlations were performed between sustained attention, as measured by the psychomotor vigilance test, and incidental and intentional memory performance for the total score and for the corrected score. Significant correlations were found only in the incidental encoding condition. Positive correlations were found between the aggregate score on vigilant attention and memory performance, with higher scores on vigilant attention correlating with higher scores on both scores of object-location memory. None of the correlations between vigilant attention and memory performance in the intentional encoding condition were significant (see Table 3). The difference between these correlations was not statistically significant ($p > 0.05$; see Figures 3 and 4).

Table 3. Correlations between vigilant attention (aggregate score) and location-exchanged objects during divided attention incidental and intentional encoding conditions.

	Incidental Encoding	Intentional Encoding
Total scores	0.30 *	0.12
Corrected scores	0.29 *	0.15

* $p < 0.05$; The total score was the total number of correct identifications in exchanged objects The corrected score was the total number of correct identifications minus the incorrect identifications in exchanged objects.

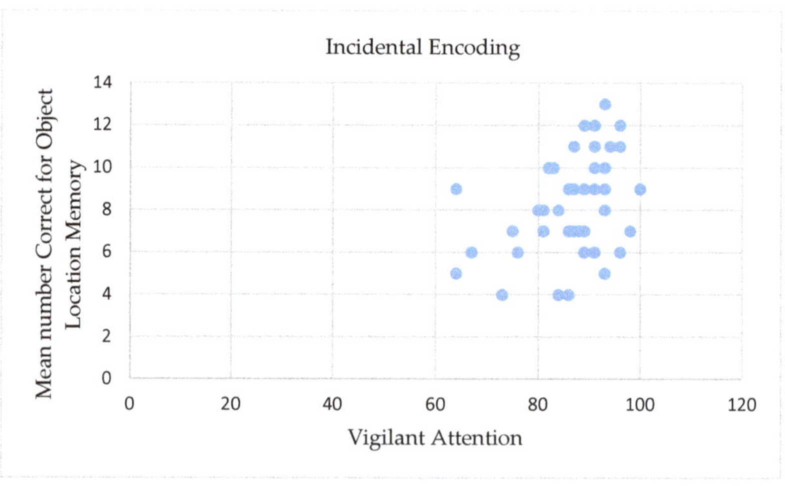

Figure 3. Mean number of correctly detected location-exchanged objects under the divided attention incidental encoding condition for the total scores as a function of vigilant attention (aggregate score).

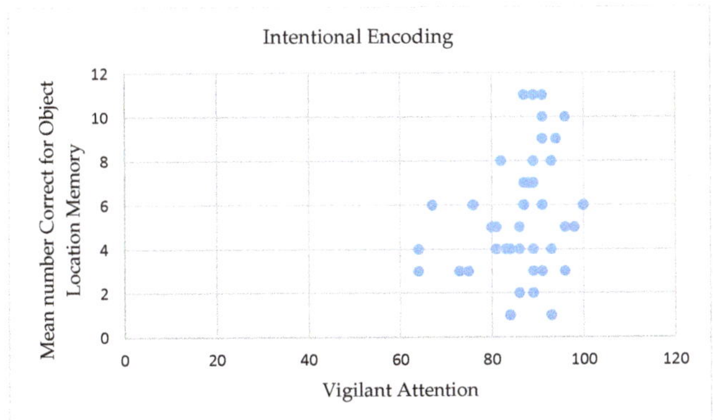

Figure 4. Mean number of correctly detected location-exchanged objects under the divided attention intentional encoding condition for the total scores as a function of vigilant attention (aggregate score).

4. Discussion

The present study aimed to examine performance differences in incidental and intentional memory under divided attention conditions. Furthermore, the present study sought to examine the relation between sustained attention and incidental and intentional memory performance. With regard to memory performance under conditions of incidental and intentional encoding, spatial memory studies are characterized by long-lasting controversies. Whereas automaticity hypothesis supporters suggest that the encoding of objects' locations can occur even without attention allocation [2], other studies find that participants' awareness of subsequent retrieval requests can improve performance (e.g., [4]). The present study provides support for the automaticity hypothesis showing that participants performed better in incidental than in intentional memory. Although previous studies that have shown that intention to learn locations improves memory performance, participants exhibited the ability to encode locations without explicit instruction to do so. The present study was conducted using a within-subject design, wherein participants responded both to the incidental condition and the intentional condition. A previous study [13] using the same paper-and-pencil format but with between-subject design, found the same results: female subjects under the incidental condition performed better than those under the intentional condition. Moreover, in the present study, participants memorized objects' locations under attention load, during which they were additionally requested to direct their attention to an auditory task. Even though their attention resources were limited due to the need to allocate attention to another task, the distraction task did not deplete their attention resources, as evidenced by their ability to encode object locations to memory in incidental conditions.

The present study additionally focuses on the relation between sustained attention and incidental and intentional memory performance. We found that sustained attention measures correlated with incidental, but not intentional, object-location memory performance. Our findings are in line with a recent study that examined the role of executive attention in relation to incidental and intentional memory, on both verbal and spatial tasks [27].

Furthermore, the present study examined memory performance under attention load conditions through a divided attention paradigm in order to uncover the role of sustained attention on wide conditions of memory encoding. Sustained attention is one of several attentional control, or executive attention, abilities [16]. Brain imaging studies implicate that several regions are associated with sustained attention, especially the anterior cingulate cortex and the right prefrontal cortex [53,54]. Additionally, Smith and colleagues [55] reported greater activity in the right hemisphere during tasks of spatial working memory, thus suggesting a coupling between memory demands and sustained attention [56]. However, the association between sustained attention and memory performance was found in the present study for the incidental encoding condition only. Based on Kontaxopoulou and colleagues' [27] finding that aging affects incidental, rather than intentional, encoding processes, they proposed that the ability to effectively execute incidental memory processes is more strongly connected with the overall cognitive system, as indicated by the association found between incidental memory and attention and executive functions. Indeed, memory studies conducted with elderly populations have shown that low scores on memory tasks were associated with reduced activation in the frontal lobes (e.g., [28]). Furthermore, imaging studies support a positive correlation between executive functioning and prefrontal cortex volume (for a meta-analysis, see [29]). Further support comes from animal studies. For example, Parnell, Grasby, and Talk [57] demonstrated that lesions on the medial prefrontal cortex impacted incidental encoding for locations in rodents. The authors suggested that the prefrontal cortex is needed for sustained attention to incidental encoding of locations.

In the realm of spatial attention, a paradigm termed *contextual cueing* [58] has been widely studied. In implicit contextual cueing tasks, repeated visual context facilitates visual search for target objects. Extensive research aimed at investigating the factors influencing spatial attention has been conducted. Researchers found that current goals, perceptual salience, statistical learning, reward, motivation and emotion affect attention [59]. Recent studies that have examined the influence of memory load on contextual cueing have presented conflicting findings. Whereas some studies have shown that contextual cueing is impaired by memory load [60], other work has shown that contextual cueing remains intact [61], suggesting that contextual cueing is sensitive to memory load under specific conditions [59].

The present study has some limitations. First, the present study focused on female participants only. Given previous results indicating sex differences in object-location memory performance and specifically the female advantage in these tasks, we chose to focus on females. However, future studies should expand the sampling frame to include male participants; this will allow for uncovering sex differences in processing strategies utilized in memory encoding and sustained attention. Second, the present study focused on divided attention conditions. The object-location memory literature has typically explored encoding manipulations under full attention conditions. Only scarcely have divided attention conditions been used [12,13]. In order to shed light on the role of sustained attention in memory encoding, the present study chose to utilize an attention load paradigm using divided attention conditions. However, to deepen our understanding on the role of attention control on memory encoding processes, a broader examination including various attention conditions (e.g., full, selective) is still needed. Third, the present study used a distraction task which did not appear to dilute the attentional resources allocated for the memory task, as illustrated by the relatively high performance of the participants. Future studies should examine various distraction modalities, including various component numbers in order to identify the precise conditions under which attentional resource allocation facilitates, as opposed to inhibits, memory performance. Fourth, although this has been customary in previous studies, the use of only one measure in the present study raises the need for replication in future studies using two measures forming sustained attention score. Fifth, due to study requirements (in the sake of face validity), all participants performed the study tasks (under incidental and intentional conditions) in the same order. Although the current findings replicate previous results found using between-subject design [13], confounding factors cannot fully be ruled out. Therefore, caution should be exercised interpreting the present findings.

5. Conclusions

In sum, the present findings demonstrated that in incidental encoding conditions, the distraction task did not completely diminish participants' attentional resources, as they exhibited high memory performance. Furthermore, the present findings suggest that memory encoding also benefits from explicit instructions to memorize locations under divided attention conditions, but to a lesser extent. Therefore, the present study supports the notion that object-location memory possesses several components that differ in processing automaticity. Furthermore, to the best of our knowledge, the present study is the first to examine the relation between sustained attention and object-location memory. We found that sustained attention plays an important role in incidental, but not in intentional, encoding, thus supporting previous findings which have examined other memory tasks. In addition, in line with the over-load theory, the present study was conducted under conditions of increased mental load. We found that under conditions of divided attention, executive attention was associated with incidental, but not with intentional encoding. These findings enable us to identify the exact conditions under which executive attention influences memory performance. Studying object-location memory through specifying the conditions by which performance declines is essential regarding aging influences on memory performance. Moreover, previous findings in animal and human imagery studies have shown that executive attention and incidental memory performance are connected with the same brain regions, including the prefrontal cortex in particular. Future studies should focus on other incidental memory tasks prominent in our daily lives and their relation to executive attention. A large body of research has demonstrated that object-location memory is even more susceptible to age-related declines than target memory [62,63]. Further examination of the role of executive attention on object-location memory, and the corresponding neural infrastructure, will shed light on the potential deleterious effects of aging on memory.

Author Contributions: Conceptualization: E.B., O.T.; Methodology: E.B., O.T.; Data analyses: E.B.; Writing: E.B., O.T.; Supervision: E.B., O.T. The authors have read and agreed to the published version of the manuscript. All authors have read and agreed to the published version of the manuscript.

Funding: This research received no external funding.

Conflicts of Interest: The authors declare no conflicts of interest.

References

1. Postma, A.; Kessels, R.P.C.; Van Asselen, M. How the brain remembers and forgets where things are: The neurocognition of object–location memory. *Neurosci. Biobehav. Rev.* **2008**, *32*, 1339–1345. [CrossRef] [PubMed]
2. Hasher, L.; Zacks, R.T. Automatic and effortful processes in memory. *J. Exp. Psychol. Gen.* **1979**, *108*, 356–388. [CrossRef]
3. Shadoin, A.L.; Ellis, N.R. Automatic processing of memory for spatial location. *Bull. Psychon. Soc.* **1992**, *30*, 55–57. [CrossRef]
4. Naveh-Benjamin, M. Recognition memory of spatial location information: Another failure to support automaticity. *Mem. Cogn.* **1988**, *16*, 437–445. [CrossRef] [PubMed]
5. Ballesteros, S.; Mayas, J. Selective attention affects conceptual object priming and recognition: A study with young and older adults. *Front. Psychol.* **2015**, *5*, 1567. [CrossRef]
6. Mulligan, N.W. The role of attention during encoding in implicit and explicit memory. *J. Exp. Psychol. Learn. Mem. Cogn.* **1998**, *24*, 27–47. [CrossRef]
7. Parkin, A.J.; Russo, R. Implicit and explicit memory and the automatid effortful distinction. *Eur. J. Cogn. Psychol.* **1990**, *2*, 71–80. [CrossRef]
8. Szymanski, K.F.; MacLeod, C.M. Manipulation of Attention at Study Affects an Explicit but Not an Implicit Test of Memory. *Conscious. Cogn.* **1996**, *5*, 165–175. [CrossRef]
9. Nussenbaum, K.; Amso, D.; Markant, J. When increasing distraction helps learning: Distractor number and content interact in their effects on memory. *Attention Perception Psychophys.* **2017**, *79*, 2606–2619. [CrossRef]
10. Chun, M. Visual working memory as visual attention sustained internally over time. *Neuro-psychology* **2011**, *49*, 1407–1409. [CrossRef]
11. Treisman, A.; Zhang, W. Location and binding in visual working memory. *Mem. Cogn.* **2006**, *34*, 1704–1719. [CrossRef] [PubMed]
12. Barel, E. The role of attentional resources in explaining sex differences in object location memory. *Int. J. Psychol.* **2016**, *53*, 365–372. [CrossRef] [PubMed]
13. Barel, E. Effects of attention during encoding on sex differences in object location memory. *Int. J. Psychol.* **2018**, *54*, 539–547. [CrossRef] [PubMed]
14. Ecuyer-Dab, I.; Robert, M. The Female Advantage in Object Location Memory According to the Foraging Hypothesis: A Critical Analysis. *Hum. Nat.* **2007**, *18*, 365–385. [CrossRef]

15. Lewin, C.; Wolgers, G.; Herlitz, A. Sex differences favoring women in verbal but not in visuospatial episodic memory. *Neuropsychology* **2001**, *15*, 165–173. [CrossRef] [PubMed]
16. Manns, J.R.; Squire, L.R. Perceptual learning, awareness, and the hippocampus. *Hippocampus* **2001**, *11*, 776–782. [CrossRef]
17. Brandt, J.; Bylsma, F.W.; Aylward, E.H.; Rothlind, J.; Gow, C.A. Impaired source memory in huntington's disease and its relation to basal ganglia atrophy. *J. Clin. Exp. Neuropsychol.* **1995**, *17*, 868–877. [CrossRef]
18. Kessels, R.P.C.; Feijen, J.; Postma, A. Implicit and Explicit Memory for Spatial Information in Alzheimer's Disease. *Dement. Geriatr. Cogn. Disord.* **2005**, *20*, 184–191. [CrossRef]
19. Unsworth, N.; Robison, M.K. Working memory capacity and sustained attention: A cognitive-energetic perspective. *J. Exp. Psychol. Learn. Mem. Cogn.* **2020**, *46*, 77–103. [CrossRef]
20. Kane, M.J.; Meier, M.E.; Smeekens, B.A.; Gross, G.M.; Chun, C.A.; Silvia, P.J.; Kwapil, T.R. Individual differences in the executive control of attention, memory, and thought, and their associations with schizotypy. *J. Exp. Psychol. Gen.* **2016**, *145*, 1017–1048. [CrossRef]
21. Langner, R.; Eickhoff, S.B. Sustaining attention to simple tasks: A meta-analytic review of the neural mechanisms of vigilant attention. *Psychol. Bull.* **2012**, *139*, 870–900. [CrossRef] [PubMed]
22. Lim, J.; Dinges, D.F. Sleep Deprivation and Vigilant Attention. *Ann. N. Y. Acad. Sci.* **2008**, *1129*, 305–322. [CrossRef] [PubMed]
23. Robertson, I.H.; O'Connell, R.G. Vigilant attention. In *Attention and Time*; Nobre, A.C., Coull, J.T., Eds.; Oxford University Press: Oxford, UK, 2010; Volume 79–88.
24. Unsworth, N.; Miller, J.D.; Lakey, C.E.; Young, D.L.; Meeks, J.T.; Campbell, W.K.; Goodie, A.S. Exploring the Relations Among Executive Functions, Fluid Intelligence, and Personality. *J. Individ. Differ.* **2009**, *30*, 194–200. [CrossRef]
25. Cheyne, J.A.; Solman, G.; Carriere, J.S.; Smilek, D. Anatomy of an error: A bidirectional state model of task engagement/disengagement and attention-related errors. *Cognition* **2009**, *111*, 98–113. [CrossRef]
26. Dockree, P.M.; Kelly, S.P.; Roche, R.; Hogan, M.J.; Reilly, R.; Robertson, I.H. Behavioural and physiological impairments of sustained attention after traumatic brain injury. *Cogn. Brain Res.* **2004**, *20*, 403–414. [CrossRef]
27. Helton, W.S.; Warm, J.S. Signal salience and the mindlessness theory of vigilance. *Acta Psychol.* **2008**, *129*, 18–25. [CrossRef]
28. Ariga, A.; Lleras, A. Brief and rare mental "breaks" keep you focused: Deactivation and reactivation of task goals preempt vigilance decrements. *Cognition* **2011**, *118*, 439–443. [CrossRef]
29. Helton, W.S.; Russell, P.N. Working memory load and the vigilance decrement. *Exp. Brain Res.* **2011**, *212*, 429–437. [CrossRef]
30. Kontaxopoulou, D.; Beratis, I.N.; Fragkiadaki, S.; Pavlou, D.; Yannis, G.; Economou, A.; Papanicolaou, A.C.; Papageorgiou, S. Incidental and Intentional Memory: Their Relation with Attention and Executive Functions. *Arch. Clin. Neuropsychol.* **2017**, *32*, 519–532. [CrossRef]
31. Rosen, A.C.; Prull, M.W.; O'hara, R.; Race, E.A.; Desmond, J.E.; Glover, G.H.; Yesavage, J.A.; Gabrieli, J.D.E. Variable effects of aging on frontal lobe contributions to memory. *NeuroReport* **2002**, *13*, 2425–2428. [CrossRef]
32. Yuan, P.; Raz, N. Prefrontal cortex and executive functions in healthy adults: A meta-analysis of structural neuroimaging studies. *Neurosci. Biobehav. Rev.* **2014**, *42*, 180–192. [CrossRef] [PubMed]
33. Naveh-Benjamin, M.; Guez, J.; Shulman, S. Older adults' associative deficit in episodic memory: Assessing the role of decline in attentional resources. *Psychon. Bull. Rev.* **2004**, *11*, 1067–1073. [CrossRef] [PubMed]
34. Peterson, D.J.; Naveh-Benjamin, M. The Role of Aging in Intra-Item and Item-Context Binding Processes in Visual Working Memory. *J. Exp. Psychol. Learn. Mem. Cogn.* **2016**, *42*, 1713–1730. [CrossRef]
35. Eals, M.; Silverman, I. The Hunter-Gatherer theory of spatial sex differences: Proximate factors mediating the female advantage in recall of object arrays. *Evol. Hum. Behav.* **1994**, *15*, 95–105. [CrossRef]
36. Healey, M.K.; Kahana, M.J. A four-component model of age-related memory change. *Psychol. Rev.* **2015**, *123*, 23–69. [CrossRef]
37. Gallagher, P.; Neave, N.; Hamilton, C.; Gray, J.M. Sex differences in object location memory: Some further methodological considerations. *Learn. Individ. Differ.* **2006**, *16*, 277–290. [CrossRef]
38. Banks, S.; Dinges, D.F. Behavioral and Physiological Consequences of Sleep Restriction. *J. Clin. Sleep Med.* **2007**, *3*, 519–528. [CrossRef]
39. Voyer, D.; Voyer, S.D.; Saint-Aubin, J. Sex differences in visual-spatial working memory: A meta-analysis. *Psychon. Bull. Rev.* **2016**, *24*, 307–334. [CrossRef]

40. Palmer, M.; Brewer, N.; Horry, R. Understanding gender bias in face recognition: Effects of divided attention at encoding. *Acta Psychol.* **2013**, *142*, 362–369. [CrossRef]
41. eWright, K.P., Jr.; Hull, J.T.; Czeisler, C.A. Relationship between alertness, performance, and body temperature in humans. *Am. J. Physiol. Integr. Comp. Physiol.* **2002**, *283*, R1370–R1377. [CrossRef]
42. Dinges, D.F.; Powell, J.W. Microcomputer analyses of performance on a portable, simple visual RT task during sustained operations. *Behav. Res. Methods Instrum. Comput.* **1985**, *17*, 652–655. [CrossRef]
43. Basner, M.; Dinges, D.F. Maximizing Sensitivity of the Psychomotor Vigilance Test (PVT) to Sleep Loss. *Sleep* **2011**, *34*, 581–591. [CrossRef] [PubMed]
44. Dinges, D.F.; Pack, A.I.; Williams, K.; Gillen, K.A.; Powell, J.W.; Ott, G.E.; Aptowicz, C. Cumulative Sleepiness, Mood Disturbance, and Psychomotor Vigilance Performance Decrements During a Week of Sleep Restricted to 4–5 Hours per Night. *Sleep* **1997**, *20*, 267–277.
45. Van Dongen, H.; Maislin, G.; Mullington, J.M.; Dinges, D.F. The cumulative cost of additional wakefulness: Dose-response effects on neurobehavioral functions and sleep physiology from chronic sleep restriction and total sleep deprivation. *Sleep* **2003**, *26*, 117–126. [CrossRef]
46. Doran, S.M.; Van Dongen, H.P.; Dinges, D.F. Sustained attention performance during sleep deprivation: Evidence of state instability. *Arch. Ital. Biol. J. Neurosci.* **2001**, *139*, 1–15.
47. Sinclair, K.L.; Ponsford, J.; Rajaratnam, S.; Anderson, C. Sustained attention following traumatic brain injury: Use of the Psychomotor Vigilance Task. *J. Clin. Exp. Neuropsychol.* **2013**, *35*, 210–224. [CrossRef] [PubMed]
48. Stuss, D.T.; Murphy, K.J.; Binns, M.A.; Alexander, M.P. Staying on the job: The frontal lobes control individual performance variability. *Brain* **2003**, *126*, 2363–2380. [CrossRef] [PubMed]
49. Johnson, M.P.; Duffy, J.F.; Dijk, D.J.; Ronda, J.M.; Dyal, C.M.; Czeisler, C.A. Short-term memory, alertness and performance: A reappraisal of their relationship to body temperature. *J. Sleep Res.* **1992**, *1*, 24–29. [CrossRef]
50. Goel, N.; Rao, H.; Durmer, J.S.; Dinges, D.F. Neurocognitive consequences of sleep deprivation. *Semin. Neurol.* **2009**, *29*, 320–339. [CrossRef]
51. Van Dongen, H.; Vitellaro, K.M.; Dinges, D.F. Individual differences in adult human sleep and wakefulness: Leitmotif for a research agenda. *Sleep* **2005**, *28*, 479–498. [CrossRef]
52. Lim, J.; Dinges, D.F. A meta-analysis of the impact of short-term sleep deprivation on cognitive variables. *Psychol. Bull.* **2010**, *136*, 375–389. [CrossRef] [PubMed]
53. Dorrian, J.; Rogers, N.L.; Dinges, D.F. Psychomotor Vigilance Performance: Neurocognitive Assay Sensitive to Sleep Loss. In *Sleep Deprivation: Clinical Issues, Pharmacology and Sleep Loss Effects*; Marcel Dekker, Inc.: New York, NY, USA, 2010; pp. 39–70.
54. Honda, A.; Nihei, Y. Sex differences in object location memory: The female advantage of immediate detection of changes. *Learn. Individ. Differ.* **2009**, *19*, 234–237. [CrossRef]
55. Smith, E.E.; Jonides, J.; Koeppe, R.A. Dissociating Verbal and Spatial Working Memory Using PET. *Cereb. Cortex* **1996**, *6*, 11–20. [CrossRef] [PubMed]
56. Parasuraman, R. Memory load and event rate control sensitivity decrements in sustained attention. *Science* **1979**, *205*, 924–927. [CrossRef] [PubMed]
57. Parnell, R.; Grasby, K.; Talk, A. The prefrontal cortex is required for incidental encoding but not recollection of source information in rodents. *Behav. Brain Res.* **2012**, *232*, 77–83. [CrossRef] [PubMed]
58. Chun, M.; Jiang, Y. Contextual Cueing: Implicit Learning and Memory of Visual Context Guides Spatial Attention. *Cogn. Psychol.* **1998**, *36*, 28–71. [CrossRef]
59. Jiang, Y.V. Habitual versus goal-driven attention. *Cortex* **2018**, *102*, 107–120. [CrossRef]
60. Travis, S.L.; Mattingley, J.B.; Dux, P.E. On the role of working memory in spatial contextual cueing. *J. Exp. Psychol. Learn. Mem. Cogn.* **2013**, *39*, 208–219. [CrossRef]
61. Vickery, T.J.; Sussman, R.S.; Jiang, Y.V. Spatial context learning survives interference from working memory load. *J. Exp. Psychol. Hum. Percept. Perform.* **2010**, *36*, 1358–1371. [CrossRef]
62. Kessels, R.P.C.; Postma, A. Object-location memory in ageing and dementia. *Adv. Consc. Res.* **2006**, *66*, 221.
63. Kessels, R.; Hobbel, D.; Postma, A. Aging, context memory and binding: A comparison of "what, where and when" in young and older adults. *Int. J. Neurosci.* **2007**, *117*, 795–810. [CrossRef] [PubMed]

Article

The Disengagement of Visual Attention: An Eye-Tracking Study of Cognitive Impairment, Ethnicity and Age

Megan Polden [1,*], Thomas D. W. Wilcockson [2] and Trevor J. Crawford [1]

1 Psychology Department, Lancaster University, Bailrigg, Lancaster LA1 4YF, UK; t.crawford@lancaster.ac.uk
2 School of Sport, Exercise and Health Sciences, Loughborough University, Epinal Way, Loughborough LE11 3TU, UK; t.wilcockson@lboro.ac.uk
* Correspondence: m.polden@lancaster.ac.uk

Received: 25 June 2020; Accepted: 14 July 2020; Published: 18 July 2020

Abstract: Various studies have shown that Alzheimer's disease (AD) is associated with an impairment of inhibitory control, although we do not have a comprehensive understanding of the associated cognitive processes. The ability to engage and disengage attention is a crucial cognitive operation of inhibitory control and can be readily investigated using the "gap effect" in a saccadic eye movement paradigm. In previous work, various demographic factors were confounded; therefore, here, we examine separately the effects of cognitive impairment in Alzheimer's disease, ethnicity/culture and age. This study included young ($N = 44$) and old ($N = 96$) European participants, AD ($N = 32$), mildly cognitively impaired participants (MCI: $N = 47$) and South Asian older adults ($N = 94$). A clear reduction in the mean reaction times was detected in all the participant groups in the gap condition compared to the overlap condition, confirming the effect. Importantly, this effect was also preserved in participants with MCI and AD. A strong effect of age was also evident, revealing a slowing in the disengagement of attention during the natural process of ageing.

Keywords: cognitive impairment; disengagement; attention; inhibition; "gap effect"; overlap; saccade

1. Introduction

Alzheimer's disease (AD) is a neurodegenerative disease that leads to profound cognitive impairment that includes changes in working memory [1,2]. AD is often diagnosed relatively late in the neuropathology of the disease, due to the lengthy and subjective assessments for the clinical diagnosis that are currently used. Subtle early impairments in executive function, attentional disengagement and other cognitive processes have been reported in people with AD [3,4]. Various attempts have been made to develop specific measures of attentional control in patients with AD [5–8]. However, these have included multiple cognitive operations or have not been grounded in neurophysiological research that has provided insights into the attentional disengagement. An exception is the work by Parasuraman and colleagues [9,10] using the Posner task. Posner [11] stated that orienting of attention comprised three distinct stages: (1) disengagement from the current stimulus; (2) movement to the new location; and (3) re-engagement with the target at the new location. According to the Posner model, attention must be disengaged from the current visual target, in order to facilitate an attentional shift from the old to the new target; just as in driving a car where you disengage from one gear, before moving the gear stick to a new gear. These distinct operations require multiple brain processes, with each contributing to the cost in terms of the overall processing time [12]. Parasuraman and colleagues [9,10] reported that the reaction times to a "valid" cue (that summoned automatic attention towards the target) was equivalent in the AD and control participants. In contrast, the reaction times to an "invalid" cue (that required disengagement of attention away from the cue) was substantially increased in the AD group. This suggested that the automatic orientation of attention was preserved in AD, but the ability to disengage attention was impaired. However, these results failed to be replicated in several laboratories [13,14].

Mounting research has demonstrated that the attentional operations used in eye tracking tasks can provide an early marker of neurodegenerative disease [15–20]. Importantly, eye movement abnormalities occur earlier than the more noticeable changes in memory, which present relatively late in the progression of the disease [21]. A dual saccadic paradigm is often used to evaluate attentional disengagement [22–25]. In the so-called "gap" condition, the fixation point is removed 200 ms prior to the presentation of the display target, resulting in a temporal "gap" between the offset of the fixation point and the presentation of the new target. This condition yields relatively fast reaction times due to the facilitation of the disengagement operation by the prior removal of the fixation point. In contrast, in the "overlap" condition, the fixation point remains for a period of time while the new target is displayed (see Figure 1a,b). Therefore, in this condition, there is a temporal overlap between the offset of the central fixation point and the onset of the target. The "gap effect" is measured by the difference in the mean saccadic reaction times between the gap and overlap conditions and yields an operational index of attentional disengagement [26,27]. A saccadic eye movement is triggered relatively early in comparison to situations where the fixation point remains visible with the peripheral target, as in the step or overlap conditions [18,25,28,29].

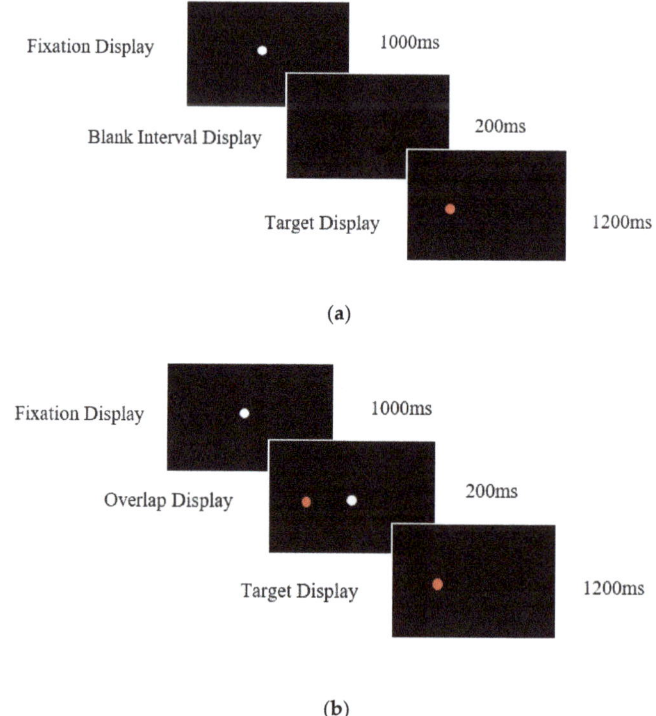

Figure 1. Gap and Overlap Displays (**a**) Timings and sequence of the prosaccade task gap condition. The gap condition facilitates the disengagement of visual attention prior to the target's presentation due to the removal of the central fixation point. (**b**) Timings and sequence of the pro-saccade task overlap condition. The central fixation point remains on for a short period when the target is displayed. This results in a delay in the disengagement of attention, resulting in longer mean saccade reaction times.

There has been relatively little research on the "gap effect" in patients with neurodegenerative disease. Prosaccades have the potential to assess attentional fluctuation in patients with neurodegenerative disease and offer an alternative to more traditional paper-based tests. The few studies that have been reported have yielded conflicting findings. For example, Yang et al. [30] reported in a sample of Chinese AD and mildly cognitively impaired (MCI) participants a substantially larger "gap effect" in comparison to healthy age-matched controls. In contrast, a recent study with Iranian participants revealed no difference in the prosaccade gap effect between AD participants and healthy controls [31]. Crawford et al. [17] found using a longitudinal design with European U.K. participants that the "gap effect" in AD was similar to that of the controls after a 12 month period. These differences could be due to a combination of methodological factors, including the participant populations, since to our knowledge, no study has contrasted different ethnicity groups within a single study design. It is important to examine the effect in various populations to determine the cultural validity of the gap effect. Restricting study populations to Western, educated, industrialized, rich and democratic (WEIRD) samples has contributed to the replicability crisis [32]. Eye movement characteristics have previously differed across ethnicity/cultural groups [33,34], therefore, comparisons across cultures is important.

In summary, this work is an exploration of attentional disengagement, to determine the potential mediating effects of: (a) cognitive impairment (contrasting European participants with AD, MCI and European healthy older participants); (b) healthy ageing (contrasting healthy young and older European participants); and (c) ethnicity/culture (contrasting older European older participants and older South Asian participants).

2. Materials and Methods

2.1. Participants

The study included 32 participants with dementia caused by Alzheimer's disease (AD: mean age = 74.32, SD = 7.57, age range = 59–86 years), 47 participants with mild cognitive impairment (MCI: mean age = 70.83, SD = 8.17, age range = 56–84 years), 96 typically ageing older European participants (mean age = 66.18, SD = 7.94, age range = 55–83 years), 44 younger European adults (mean age = 21.13, SD = 2.87, age range = 18–26 years), and 94 South Asian older adults (mean age = 67.25, SD = 6.13, age range = 55–79 years). Older and younger European participants were white British or European fluent English speakers with a minimum of 11 years in formal education. The older European participants were recruited from the local community, with the younger adults recruited via Lancaster University's Research Participant System. The Asian participants were recruited from local Hindu temples located in the northwest of England, who were born in India or East Africa, but had resided in the U.K. for an average of 46.66 years (SD = 5.94).

The AD and MCI participants were recruited via various NHS sites and memory clinics across the U.K. Participants had received a clinical diagnosis following a full assessment from a dementia specialist. AD participants had a formal diagnosis of dementia due to AD and met the requirements for the American Psychiatric Association's Diagnostic and Statistical Manual of Mental Disorders (DSM IV) and the National Institute of Neurological and Communicative Disorders and Stroke (NINCDS) for AD. MCI participants met the following criteria [35] and had a diagnosis of dementia due to mild cognitive impairment: (1) subjective reports of memory decline (reported by the individual or caregiver/informant); (2) memory and/or cognitive impairment (scores on standard cognitive tests were >1.5 SDs below age norms); (3) activities of daily living were preserved. The following exclusion criteria were applied: patients with acute physical symptoms, focal cerebral lesions, history of neurological disease (e.g., Parkinson's disease, multiple sclerosis, epilepsy, amyotrophic lateral sclerosis, muscular dystrophy), cerebrovascular disorders (including ischemic stroke, haemorrhagic stroke, atherosclerosis), psychosis, active or past alcohol or substance misuse/dependence or any physical or mental condition severe enough to interfere with their ability to participate in the study.

All participants retained the capacity to consent to participation in the study and provided written informed consent. Ethical approval was granted by the Lancaster University Ethics committee and by the NHS Health Research Authority, Greater Manchester West Research Ethics Committee.

2.2. Neuropsychological Assessments

The Montreal Cognitive Assessment (MoCA) [36] was administered as an indicator of probable dementia with a score of 26/30 or higher considered normal. The digit [37] and spatial span [37], forward and reverse, were used to estimate short-term memory span and working memory.

2.3. Eye Tracking Tasks

2.3.1. Apparatus

Eye movements were recorded using SR EyeLink Desktop 1000 with a sampling rate of 500 Hz. A chin rest was used to minimise head movements, and participants were seated 55 cm away from the computer screen. Prior to the start of each eye tracking task, a 9 point calibration was used. The stimulus was controlled and created via the use of Experiment Builder Software Version 1.10.1630.

2.3.2. Prosaccade Task

Participants were presented with 36 gap trials followed by 12 overlap trials. A white central fixation point was displayed for 1000 ms, followed by a red target presented randomly at 4° to the left or right for 1200 ms. Participants were instructed to look towards the central fixation point and then when the red target appeared to move their gaze towards it as quickly and accurately as possible. Between trials, a black interval screen was displayed for 3500 ms.

The gap condition included a blank interval screen displayed for 200 ms between the initial appearance of the red target and the extinguishment of the central fixation target. For this condition, the red and white target never appeared on the screen simultaneously (Figure 1). In the overlap condition, the target was presented while the central fixation remained present on the screen for a short period. There was a 200 ms "overlap" in which the target and fixation point were presented simultaneously (Figure 2). After this period, the central fixation was removed, and the target presented singularly for 1200 ms. Previous research [16,18] found that this format works well for patients with neurodegenerative diseases.

2.4. Data Analysis

The raw data were analysed and extracted from the EyeLink using DataViewer Software Version 3.2. The raw data were analysed offline via the use of the software [38]. The software filtered noise and spikes by removing frames with a velocity signal greater than 1500 deg/s or an acceleration signal greater than 100,000 deg^2/s. The fixations and saccadic events were detected by the EyeLink Parser, and the saccades were extracted alongside multiple spatial and temporal variables. Trials in which the participants did not direct their gaze to the fixation point before the target display were removed. Anticipatory saccades made prior to 80 ms and excessively delayed saccades over 700 ms were also filtered from the data.

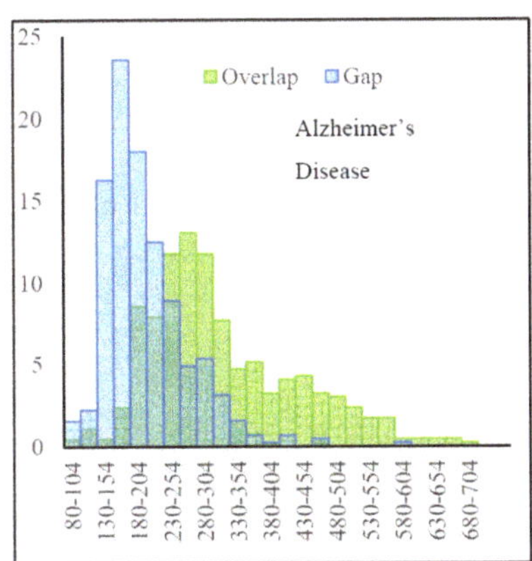

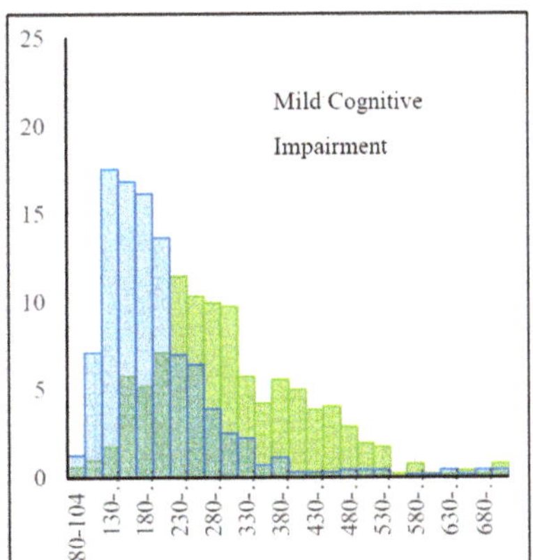

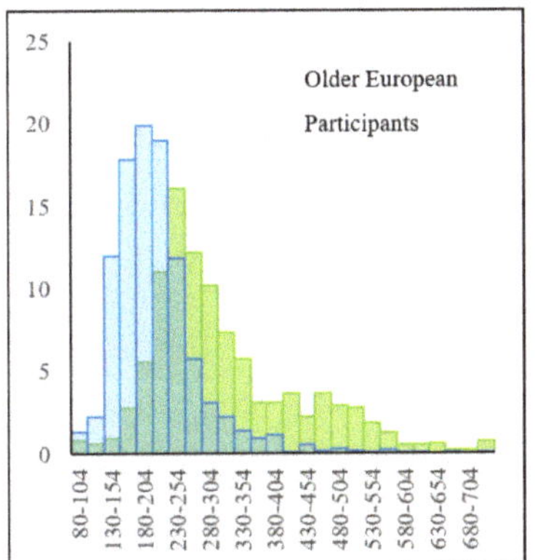

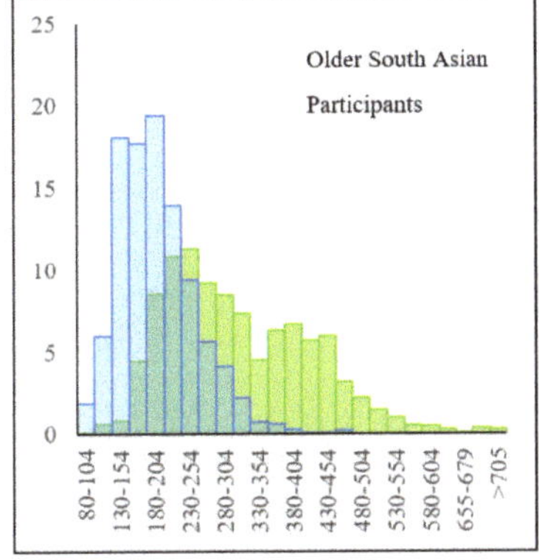

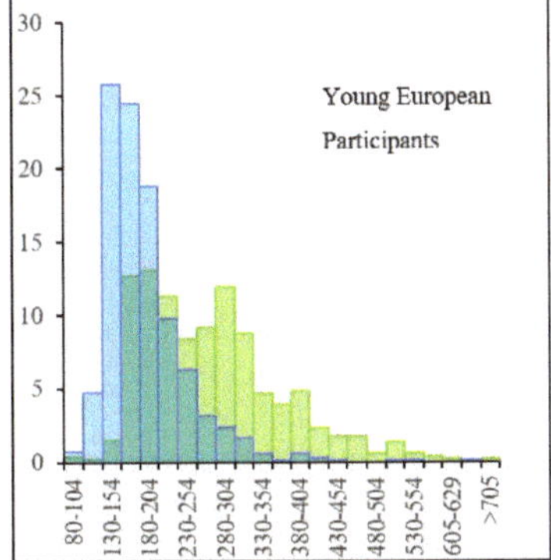

Saccade Latency (ms)

Figure 2. Histograms displaying a shift in the distribution of saccade latencies in the gap condition (blue) compared to the overlap condition (green) for the participant groups: Alzheimer's disease, mild cognitive impairment, older and younger European participants and older South Asian participants.

3. Results

Linear mixed effects model analyses were carried out using RStudio Version 1.2.5033. The models conducted an analysis of the reaction times in the gap and overlap conditions. The "gap effect" value was calculated by subtracting the individuals mean latency in the gap condition from the overlap condition mean latency. The linear mixed effects model also determined the group effects of disease, ageing, and ethnicity. Two participants in the MCI group were excluded from the subsequent analyses due to their mean reaction times in the prosaccade gap and overlap condition being greater than two standard deviations away from the mean.

3.1. Neuropsychological Tests

A linear mixed effects model was conducted to analyse the performance on the neuropsychological tests. Table 1 shows that there was the expected effect of disease on the MoCA test, with lower scores for the AD participants compared to older European participants, $\beta = 6.90$, $t\ (257) = 7.25$, $p < 0.0001$. AD participants scored significantly lower than the MCI participants, $\beta = 2.83$, $t\ (257) = 2.72$, $p = 0.007$ (see Table 2). The European older adults produced higher MoCA scores than the MCI group ($\beta = -4.06$, $t\ (257) = -5.13$, $p < 0.001$) and unexpectedly the South Asian group ($\beta = -6.89$, $t\ (257) = -7.25$, $p < 0.001$). This difference could be due to the combination of culturally inappropriate test items, linguistics and other cultural factors. There were no significant differences in the MoCA between the healthy European older and younger adults.

Table 1. Table displaying mean reaction times and standard deviations for the neurological assessments. MCI—mild cognitive impairment.

	Older European Participants		Older South Asian Participants		Alzheimer's Disease		MCI		Young European Participants	
	M	SD	M	SD	M	SD	M	SD	M	SD
MoCA	27.80	2.04	22.04	4.99	20.19	5.45	22.98	5.40	28.14	1.94
Digit Span Task	17.91	4.60	13.27	3.71	15.23	4.56	15.95	4.12	19.86	4.33
Spatial Span Task	13.89	2.44	12.47	2.24	11.42	3.75	12.93	3.08	17.38	2.08

Note: dependent variable: total task score.

Table 2. Table displaying post hoc comparisons for the neurological assessments.

	Post Hoc Contracts (p Values)				
	Disease Effects			Ageing Effects	Ethnicity Effects
	AD vs. OEP	AD vs. MCI	MCI vs. OEP	OEP vs. YEP	OEP vs. OSP
MoCA	<0.001 *	0.007 *	<0.001 *	0.025	<0.001 *
Digit Span Task	0.015 *	0.496	0.043	0.015 *	<0.001 *
Spatial Span Task	<0.001 *	0.021 *	0.077	<0.001 *	0.002 *

AD—Alzheimer's disease; MCI—mild cognitive impairment; OEP—older European participants; OSP—older South Asian participants. YEP—young European participants. * Significant at $p < 0.05$.

The digit span test (total score forward and backward) revealed that AD participants had a significantly lower mean score than the older European participants, $\beta = 2.38$, $t\ (254) = 2.46$, $p = 0.015$. No significant differences were found between the AD and MCI group (see Table 2). The older South Asian participants had significantly lower digit span than the older European participants ($\beta = -4.45$, $t\ (254) = -6.49$, $p < 0.001$). A significant difference was found between younger and older European adults, with a higher mean digit span score for the younger adults ($\beta = 2.25$, $t\ (254) = 2.45$, $p = 0.015$).

Table 1 shows the results from the spatial span (forward and backward) and revealed that, as expected, the AD participants scored significantly lower compared to the older European participants, $\beta = 2.41$, $t\ (242) = 3.99$, $p < 0.001$. The AD participants had a significantly lower spatial span score than MCI participants, $\beta = 1.50$, $t\ (242) = 2.32$, $p = 0.021$. The findings revealed an ageing effect with young adults producing significantly higher spatial span scores than the European older adults ($\beta = 3.53$, $t\ (242) = 6.15$, $p < 0.001$). The older South Asian participants scored significantly lower than older European participants, $\beta = -1.39$, $t\ (242) = -3.17$, $p = 0.002$ (Table 2).

3.2. The "Gap" Effect

Figure 2 shows the relative shift in the latency distributions in the gap and overlap trials for each of the participant groups. A linear mixed model analysis was conducted to analyse the reaction times in relation to the participant groups. The overlap condition yielded significantly longer reaction times overall, compared to the gap condition, $\beta = 108.21$, $t\ (8881) = 57.33$, $p < 0.0001$ (Figure 2). The "gap effect" was therefore evident in all groups, with significantly faster reaction times in the gap condition, compared to the overlap condition.

3.3. Attentional Disengagement: Effects of Ageing

The older European participants' and the younger European participants' reaction times were compared in the gap and overlap conditions to determine the effects of age. Table 3 reveals that the mean "gap effect" was significantly smaller in the younger European participants (87 ms) compared to the older European participants (110 ms) $\beta = -23.46$, $t\ (315) = -2.31$, $p = 0.022$. Results showed baseline differences in prosaccades with younger European participants having significantly faster reaction times in the gap ($\beta = -8.22$, $t\ (4624) = -2.70$, $p = 0.007$) and overlap conditions ($\beta = -38.46$, $t\ (4257) = -6.81$, $p < 0.001$) compared to older European participants. This indicated that older European participants showed a greater difficulty in disengaging attention from the central fixation in comparison to the younger adults in addition to a general slowing in prosaccades.

Table 3. Table displaying mean reaction times and standard deviations for the prosaccade task gap and overlap conditions.

	Older European Participants N = 96		Older South Asian Participants N = 94		Alzheimer's Disease N = 32		MCI N = 45		Young European Participants N = 44	
	M	SD	M	SD	M	SD	M	SD	M	SD
Gap	195	38.87	212	37.06	206	30.93	200	42.18	185	31.60
Overlap	305	75.06	315	75.06	312	51.32	310	66.86	272	58.83
Gap Effect (ms) (Overlap-Gap)	110	57.30	103	58.66	106	48.06	110	59.54	87	48.53

3.4. Attentional Disengagement: Effects of Cognitive Impairment

Table 4 reveals that there was a significant difference between the AD and older European participants' saccadic reaction times in the gap condition ($\beta = -10.20$, $t\ (4624) = -2.92$, $p = 0.004$). There was no significant difference in reaction times in the overlap ($\beta = -2.41$, $t\ (4257) = -0.361$, $p = 0.718$) condition. There was also no significant difference between the "gap effect" between the conditions ($\beta = -4.29$, $t\ (315) = -0.376$, $p = 0.707$). Similarly, there were no significant differences in reaction times in these conditions when comparing the AD group with the MCI group (Table 4). There were no significant differences between the MCI and European older controls in the overlap condition; however, in the gap condition, MCI participants revealed a significant increase in mean saccadic reaction times compared to the European older controls. Thus, prosaccades and the "gap effect" were generally well preserved in people with AD and MCI.

3.5. Attentional Disengagement: Ethnicity/Cultural Effects

The older European group was contrasted with the South Asian older adults to determine the effects of ethnicity on prosaccade reaction times and the "gap effect". The results shown in Table 4 revealed that the European older group generated faster reaction times compared to the South Asian older group ($\beta = 15.78$, $t\ (4624) = 6.28$, $p < 0.001$) in the gap condition and overlap conditions ($\beta = 9.95$,

t (4257) = 2.42, p = 0.016). There was no difference in the proportion of the "gap effect" between the groups (Table 4).

Table 4. Table displaying post hoc comparisons for the prosaccade task gap and overlap conditions.

	Post Hoc Contracts (p Values)				
	Disease Effects			Ageing Effects	Ethnicity Effects
	AD vs. OEP	AD vs. MCI	MCI vs. OEP	OEP vs. YEP	OEP vs. OSP
Gap	0.004 *	0.161	<0.001 *	0.007 *	<0.001 *
Overlap	0.718	0.972	0.706	<0.001 *	0.016 *
Gap Effect (Overlap-Gap)	0.707	0.885	0.803	0.022 *	0.383

AD—Alzheimer's disease; MCI—mild cognitive impairment; OEP—older European participants; OSP—older South Asian participants. YEP—young European participants. * Significant at $p < 0.05$.

3.6. Correlations

The neuropsychological measures of memory yielded separate scores: forward, backward and total scores for digit and spatial memory, thus six measures of memory in total. The forward recall score yielded an index for memory span, whilst the backward recall score yielded a more direct measure of working memory, since it relied not simply on pure recall, but also cognitive manipulation of the items in short-term memory. Table 5 reveals that there was a significant negative correlation for backward spatial memory and the gap-effect for the MCI group, such that people with longer attentional disengagement reactions times were associated with lower spatial working memory. Interestingly, this relationship was not evident for digit span, which probed verbal working memory. Curiously, this relationship appeared to be specific to the MCI group, although it was not clear why this relationship was specific to MCI. For many participants, MCI was an intermediate transition state between healthy cognition and Alzheimer's disease. A significant proportion, but by no means all, will unfortunately go on to develop full-blown AD, although we do not yet have a reliable predictive behavioural measure of those people with MCI who will progress to AD. It appears that during this transition period, attentional disengagement may provide a useful index of the decline in working memory, and the progression from MCI to AD. Longitudinal studies will be required to determine the validity of this hypothesis.

Table 5. Correlations of prosaccade conditions and neuropsychological tests.

Variable		MoCA	Digit Span Total	Digit Span Forward	Digit Span Backward	Spatial Span Total	Spatial Span Forward	Spatial Span Backward
Gap Effect (Overlap-Gap)	AD	−0.063	0.120	0.048	0.169	0.157	0.242	0.058
	MCI	−0.096	−0.076	−0.118	−0.015	−0.168	0.021	−0.318 *
	OEP	−0.095	−0.004	−0.014	−0.031	−0.014	0.025	−0.045
	OSP	0.213	0.153	0.144	0.126	−0.024	−0.028	−0.013
	YEP	0.147	−0.013	0.070	−0.088	0.074	0.071	0.043

AD—Alzheimer's disease; MCI—mild cognitive impairment; OEP—older European participants; OSP—older South Asian participants. YEP—young European participants. * Significant at $p < 0.05$.

4. Discussion

This study revealed that the "gap effect" was well preserved in AD and MCI participants. Participants produced significantly faster reaction times when performing pro-saccadic eye movements during the gap condition compared to the overlap condition. Moreover, the effect was robust across both ethnic/cultural groups explored in this study.

4.1. What Does the Gap Effect Reveal about the Integrity of the Alzheimer Brain?

The neurophysiological networks that regulate the control of saccadic eye movements are relatively well understood. The saccadic eye movements are generated by precise reciprocal activation of saccade-related neurons and the inhibition of fixation neurons in the superior colliculus [39,40].

According to the Findlay and Walker [41] model, the removal of the fixation target leads to a reduction in the activation of the fixation units, which releases the saccade from inhibition, and this is reflected by the reduction in reaction times. When the fixation point remains on, the fixation units are tonically active, and the move units are inhibited, causing a delay in the initiation of a saccade. This network is clearly well preserved in early and late stages of the disorder. In previous work, we examined inhibitory control saccades extensively using the antisaccade task. In contrast to the gap and overlap task, the anti-saccade task requires that the observer looks away from the object, in the opposite direction, and is one of the most widely used paradigms assessing inhibitory control in both healthy individuals and clinical disorders [42,43]. These studies have shown that people with dementia generate a high proportion of uncorrected prosaccade errors towards the target in the antisaccade task that correlates with the severity of the dementia [16]. In contrast, when healthy participants make errors, they are normally rapidly corrected, although both AD and MCI adhere to the principle that the frequency of past errors predicts the probability of future errors [44]. People with amnesic MCI are at a greater risk of progressing to dementia [45–47]. Recently, our lab has shown that these errors are also evident in amnesic MCI to a greater extent than non-amnesic MCI participants [20]. We have argued that this error correction implicates a neural network that includes the anterior cingulate. Together with this work, the current evidence of the preservation of the attentional disengagement [16] will help to increase our understanding of the specificity of oculomotor impairment in AD and undermine the idea that the source of the uncorrected errors can be attributed to the inability to disengage attention from the prepotent target. Rather, the inhibition appears to be directly linked to top-down inhibitory control and working memory [16,17]. Clearly, there is a dissociation of impairment of the oculomotor pathways in AD. Evidence from this study revealed a preservation of the superior colliculus pathway, while converging evidence from previous and more recent work [20] indicated that other centres of the network, including the anterior cingulate, that mediate top-down inhibitory control and error monitoring are affected early in the course of the disease [21].

4.2. Ageing

Another key finding was a strong ageing effect on the saccadic reaction times. Although all participant groups displayed the gap effect, the younger adults revealed a significantly faster mean reaction times in the overlap and gap conditions than the older adults. Previous research has reported that eye movements are susceptible to ageing effects, in particular reductions in processing speed, spatial memory and inhibitory control [48–51]. Crawford et al. [17] reported that the "gap effect" increased in older adults compared to younger adults, suggesting that the changes in the attentional engagement are associated with normal ageing, rather than AD. The older adults are apparently more dependent on the removal of the central stimulus to facilitate the shift of attention from fixation and therefore showed a larger benefit following the removal of the fixation point in the gap condition compared to the younger adults. One possible explanation is that may be due to an age-related decrease in the reciprocal inhibitory activity of the fixation and move units [41].

4.3. Ethnicity

As outlined above, the European and the South Asian older adults both demonstrated the gap effect, with significantly faster prosaccades in the gap condition compared to the overlap condition. Clear differences between the groups emerged in the saccade reaction times, specifically for the saccade gap and overlap conditions, with South Asian adults presenting slower saccade reaction times. This raises the possibility that the south Asian group may have a lower proportion of fast and express saccades in the gap and overlap tasks. Express saccades [52,53] are fast reaction time saccades (80 ms–130 ms) with frequencies that vary across cultural groups. The frequency of express saccades was reduced in the overlap task, because the temporal overlap of the fixation-point and the target often inhibited the prosaccade, which may have reduced the difference in the overlap condition. Knox, Amatya, Jiang and Gong [33] demonstrated that Chinese participants showed a higher proportion of express saccades compared to U.K. participants. Clearly, saccade performance can differ across different cultural groups. If express saccades were a contributory factor to the faster saccade latencies of the European group in the gap condition, this would explain the convergence of saccade latencies for the groups in the overlap condition. However, the combined group latency distributions in Figure 2 suggests that this hypothesis may be flawed and cannot account for the group differences in the gap task and overlap task, although this would be best examined within a design with a larger number of trials, with the distributions of individual participants.

Previous research has also shown differences in eye movements across different cultural and ethnic groups [33,54,55]. Chua, Boland and Nisbett [56] found differences in scan patterns between native Chinese and native English-speaking participants when assessing visual scenes. English participants tended to look first at the foreground object and had an increased number of fixations then Chinese participants, predominantly focused on the background visual areas of the scene. Eye movements are clearly not homogeneous; culture and ethnicity factors can influence specific features of eye movement control.

Knox and Wolohan [57] examined saccades in European, Chinese and U.K.-born Chinese participants who shared similar cultural experiences as the European group. The study investigated whether the differences in the saccadic eye movements of the Chinese and European groups resulted from cultural or culture-unrelated factors. The Chinese participants showed similar pattern results irrespective of the culture exposure. Therefore, cultural differences cannot be the primary cause of the difference in oculomotor characteristics. Although the principal explanatory factors of these differences in oculomotor systems is unclear, they are possibly related to a combination of genetic, epigenetic and environmental factors [58]. A recent study showing very clear differences in the post-saccadic oscillations of Chinese-born and U.K.-born undergraduates concluded that "..genetic, racial, biological, and/or cultural differences can affect the morphology of the eye movement data recorded and should be considered when studying eye movements and oculomotor fixation and saccadic behaviors" [34]. Although there has been increasing eye-tracking research with Chinese participants, research with South Asian populations has been sparse. We hope that this work will encourage future studies to help redress this void.

5. Conclusions

Research scientists have tended to focus on the memory, intelligence and other mental skills that degenerate in AD and understandably have paid less attention to those equally important cognitive functions that may be well preserved. A better understanding of preserved functions in the disease will help to develop potential new early intervention strategies in the treatment of the disease that may improve mental functions and delay the progression of the disease. Patients with AD show large individual differences in the profile of scores across both traditional cognitive assessment and measures of saccadic eye movement. Therefore, in our recent work [18,21], we developed a profile measure of z-scores for each test that captures a patient's performance across a range of measures in relation to the normative scores. This approach takes full advantage of the extensive range of saccadic eye movement parameters to assess and monitor cognitive changes in the evolution of AD and will enhance specificity and sensitivity as a diagnostic tool.

Further, this study demonstrated that prosaccades can be susceptible to disease, ageing and ethnicity effects, and therefore, future research should strive to include non-WEIRD participant groups to create a more comprehensive understanding of the effect and its robustness and generalizability.

Author Contributions: Conceptualization, methodology, software, validation, investigation, writing, review and editing, contributions and project administration contributions were made by T.J.C., T.D.W.W. and M.P. The formal analysis, data curation, writing, original draft preparation contributions were made by T.J.C. and M.P. Supervision and funding acquisition were conducted by T.J.C. All authors read and agreed to the published version of the manuscript.

Funding: This research was funded by EPSRC Grant EP/M006255/1 and the Sir John Fisher Foundation.

Acknowledgments: We are grateful to Diako Mardanbegi for the help with the data processing and to Nadia Maalin, Alex Devereaux and Claire Kelly for their help with data collection. We would like to thank the NHS researchers for their continued support and help with recruitment and data collection.

Conflicts of Interest: The authors declare no conflict of interest.

References

1. Kaufman, L.D.; Pratt, J.; Levine, B.; Black, S.E. Antisaccades: A probe into the dorsolateral prefrontal cortex in Alzheimer's disease. A critical review. *J. Alzheimer's Dis.* **2010**, *19*, 781–793. [CrossRef]
2. Belleville, S.; Bherer, L.; Lepage, É.; Chertkow, H.; Gauthier, S. Task switching capacities in persons with Alzheimer's disease and mild cognitive impairment. *Neuropsychologia* **2008**, *46*, 2225–2233. [CrossRef] [PubMed]
3. Baddeley, A.; Chincotta, D.; Adlam, A. Working memory and the control of action: Evidence from task switching. *J. Exp. Psychol. Gen.* **2001**, *130*, 641. [CrossRef]
4. Solfrizzi, V.; Panza, F.; Torres, F.; Capurso, C.; D'Introno, A.; Colacicco, A.M.; Capurso, A. Selective attention skills in differentiating between Alzheimer's disease and normal aging. *J. Geriatr. Psychiatry Neurol.* **2002**, *15*, 99–109. [CrossRef] [PubMed]
5. Awh, E.; Jonides, J. Spatial working memory and spatial selective. In *Attention in The Attentive Brain*; Parasuraman, R., Ed.; MIT Press: Cambridge, MA, USA, 1998; pp. 353–380.
6. Della Sala, S.; Laiacona, M.; Spinnler, H.; Ubezio, C.A. A Cancellation test: Its reliability in assessing attentional deficits in Alzheimer's disease. *Psychol. Med.* **1992**, *22*, 885–901. [CrossRef] [PubMed]
7. Baddeley, A.; Baddeley, H.A.; Bucks, R.S.; Wilcock, G.K. Attentional control in Alzheimer's disease. *Brain* **2001**, *124*, 1492–1508. [CrossRef] [PubMed]
8. Perry, R.J.; Hodges, J.R. Attention and executive deficits in Alzheimer's Disease; a critical review. *Brain* **1999**, *122*, 383–404. [CrossRef]
9. Parasuraman, R.; Haxby, J.V. Attention and brain function in Alzheimer's disease: A review. *Neuropsychology* **1993**, *7*, 242–272. [CrossRef]
10. Parasuraman, R.; Greenwood, P.M. Selective attention in aging and dementia. In *The Attentive Brain*; Parasuraman, R., Ed.; MIT Press: Cambridge, MA, USA, 1998; pp. 461–487.
11. Posner, M.I.; Cohen, Y.; Rafal, R.D. Neural systems control of spatial orienting. *Philos. Trans. R. Soc. Lond. Ser. B Biol. Sci.* **1982**, *298*, 187–198.
12. Posner, M.I.; Petersen, S.E. The attention system of the human brain. *Annu. Rev. Neurosci.* **1990**, *13*, 25–42. [CrossRef]
13. Caffarra, P.; Riggio, L.; Malvezzi, L.; Scaglioni, A.; Freedman, M. Orienting of visual attention in Alzheimer's disease: Its implication in favor of the interhemispheric balance. *Neuropsychiatry Neuropsychol. Behav. Neurol.* **1997**, *10*, 90–95. [PubMed]
14. Faust, M.E.; Balota, D.A. Inhibition of return and visuospatial attention in healthy older adults and individuals with dementia of the Alzheimer type. *Neuropsychology* **1997**, *11*, 13–29. [CrossRef] [PubMed]
15. Boxer, A.L.; Garbutt, S.; Rankin, K.P.; Hellmuth, J.; Neuhaus, J.; Miller, B.L.; Lisberger, S.G. Medial versus lateral frontal lobe contributions to voluntary saccade control as revealed by the study of patients with frontal lobe degeneration. *J. Neurosci.* **2006**, *26*, 6354–6363. [CrossRef] [PubMed]
16. Crawford, T.J.; Higham, S.; Renvoize, T.; Patel, J.; Dale, M.; Suriya, A.; Tetley, S. Inhibitory control of saccadic eye movements and cognitive impairment in Alzheimer's disease. *Biol. Psychiatry* **2005**, *57*, 1052–1060. [CrossRef]
17. Crawford, T.J.; Higham, S.; Mayes, J.; Dale, M.; Shaunak, S.; Lekwuwa, G. The role of working memory and attentional disengagement on inhibitory control: Effects of aging and Alzheimer's disease. *Age* **2013**, *35*, 1637–1650. [CrossRef] [PubMed]
18. Crawford, T.J. The disengagement of visual attention in Alzheimer's disease: A longitudinal eye-tracking study. *Front. Aging Neurosci.* **2015**, *7*, 118. [CrossRef]
19. Kaufman, L.D.; Pratt, J.; Levine, B.; Black, S.E. Executive deficits detected in mild Alzheimer's disease using the antisaccade task. *Brain Behav.* **2012**, *2*, 15–21. [CrossRef] [PubMed]

20. Wilcockson, T.D.; Mardanbegi, D.; Xia, B.; Taylor, S.; Sawyer, P.; Gellersen, H.W.; Crawford, T.J. Abnormalities of saccadic eye movements in dementia due to Alzheimer's disease and mild cognitive impairment. *Aging* **2019**, *2019 11*, 5389.
21. Crawford, T.J.; Higham, S. Distinguishing between impairments of working memory and inhibitory control in cases of early dementia. *Neuropsychologia* **2016**, *81*, 61–67. [CrossRef]
22. Fischer, B.; Boch, R. Saccadic eye movements after extremely short reaction times in the monkey. *Brain Res.* **1983**, *260*, 21–26. [CrossRef]
23. Fischer, B.; Weber, H. Express saccades and visual attention. *Behav. Brain Sci.* **1993**, *16*, 553–567. [CrossRef]
24. Goldring, J.; Fischer, B. Reaction times of vertical prosaccades and antisaccades in gap and overlap tasks. *Exp. Brain Res.* **1997**, *113*, 88–103. [CrossRef] [PubMed]
25. Crawford, T.J.; Parker, E.; Solis-Trapala, I.; Mayes, J. Is the relationship of prosaccade reaction times and antisaccade errors mediated by working memory? *Exp. Brain Res.* **2011**, *208*, 385–397. [CrossRef] [PubMed]
26. Saslow, M.G. Effects of components of displacement-step stimuli upon latency for saccadic eye movement. *Josa* **1967**, *57*, 1024–1029. [CrossRef]
27. Rolfs, M.; Vitu, F. On the limited role of target onset in the gap task: Support for the motor-preparation hypothesis. *J. Vis.* **2007**, *7*, 7. [CrossRef]
28. Vernet, M.; Yang, Q.; Gruselle, M.; Trams, M.; Kapoula, Z. Switching between gap and overlap pro-saccades: Cost or benefit? *Exp. Brain Res.* **2009**, *197*, 49. [CrossRef]
29. Kapoula, Z.; Yang, Q.; Vernet, M.; Dieudonné, B.; Greffard, S.; Verny, M. Spread deficits in initiation, speed and accuracy of horizontal and vertical automatic saccades in dementia with Lewy bodies. *Front. Neurol.* **2010**, *1*, 138. [CrossRef]
30. Yang, Q.; Wang, T.; Su, N.; Xiao, S.; Kapoula, Z. Specific saccade deficits in patients with Alzheimer's disease at mild to moderate stage and in patients with amnestic mild cognitive impairment. *Age* **2013**, *35*, 1287–1298. [CrossRef]
31. Chehrehnegar, N.; Nejati, V.; Shati, M.; Esmaeili, M.; Rezvani, Z.; Haghi, M.; Foroughan, M. Behavioral and cognitive markers of mild cognitive impairment: Diagnostic value of saccadic eye movements and Simon task. *Aging Clin. Exp. Res.* **2019**, *31*, 1591–1600. [CrossRef]
32. Rad, M.S.; Martingano, A.J.; Ginges, J. Toward a psychology of Homo sapiens: Making psychological science more representative of the human population. *Proc. Natl. Acad. Sci. USA* **2018**, *115*, 11401–11405. [CrossRef]
33. Knox, P.C.; Amatya, N.; Jiang, X.; Gong, Q. Performance deficits in a voluntary saccade task in Chinese "express saccade makers". *PLoS ONE* **2012**, *7*, e47688. [CrossRef]
34. Mardanbegi, D.; Wilcockson, T.D.W.; Killick, R.; Xia, B.; Gellersen, H.; Sawyer, P.; Crawford, T.J. A comparison of post-saccadic oscillations in European-Born and China-Born British University Undergraduates. *PLoS ONE* **2020**, *15*, e0229177. [CrossRef] [PubMed]
35. Lenoble, Q.; Bubbico, G.; Szaffarczyk, S.; Pasquier, F.; Boucart, M. Scene categorization in Alzheimer's disease: A saccadic choice task. *Dement. Geriatr. Cogn. Disord. Extra* **2015**, *5*, 1–12. [CrossRef] [PubMed]
36. Nasreddine, Z.S.; Phillips, N.A.; Bédirian, V.; Charbonneau, S.; Whitehead, V.; Collin, I.; Chertkow, H. The Montreal Cognitive Assessment, MoCA: A brief screening tool for mild cognitive impairment. *J. Am. Geriatr. Soc.* **2005**, *53*, 695–699. [CrossRef]
37. Wechsler, D. *WAIS-III: Administration and Scoring Manual: Wechsler Adult Intelligence Scale*, 3rd ed.; Psychological Corporation: New York, NY, USA, 1997.
38. Mardanbegi, D.; Wilcockson, T.; Sawyer, P.; Gellersen, H.; Crawford, T. SaccadeMachine: Software for analyzing saccade tests (anti-saccade and pro-saccade). In Proceedings of the 11th ACM Symposium on Eye Tracking Research & Applications, Denver, CO, USA, 25–28 June 2019; pp. 1–8.
39. Dorris, M.C.; Munoz, D.P. A neural correlate for the gap effect on saccadic reaction times in monkey. *J. Neurophysiol.* **1995**, *73*, 2558–2562. [CrossRef]
40. Dorris, M.C.; Pare, M.; Munoz, D.P. Neuronal activity in monkey superior colliculus related to the initiation of saccadic eye movements. *J. Neurosci.* **1997**, *17*, 8566–8579. [CrossRef]
41. Findlay, J.M.; Walker, R. A model of saccade generation based on parallel processing and competitive inhibition. *Behav. Brain Sci.* **1999**, *22*, 661–674. [CrossRef]
42. Hutton, S.B.; Ettinger, U. The antisaccade task as a research tool in psychopathology: A critical review. *Psychophysiology* **2006**, *43*, 302–313. [CrossRef]

43. Broerse, A.; Crawford, T.J.; Den Boer, J.A. Parsing cognition in schizophrenia using saccadic eye movements: A selective overview. *Neuropsychologia* **2001**, *39*, 742–756. [CrossRef]
44. Crawford, T.; Taylor, S.; Mardanbegi, D.; Polden, M.; Wilcockson, T.; Killick, R.; Sawyer, P.; Gellersen, H.; Leroi, I. The Effects of Previous Error and Success in Alzheimer's Disease and Mild Cognitive Impairment. *Sci. Rep.* **2019**, 9. [CrossRef]
45. Fischer, P.; Jungwirth, S.; Zehetmayer, S.; Weissgram, S.; Hoenigschnabl, S.; Gelpi, E.; Krampla, W.; Tragl, K.H. Conversion from subtypes of mild cognitive impairment to Alzheimer dementia. *Neurology* **2007**, *68*, 288–291. [CrossRef] [PubMed]
46. Yaffe, K.; Petersen, R.C.; Lindquist, K.; Kramer, J.; Miller, B. Subtype of mild cognitive impairment and progression to dementia and death. *Dement. Geriatr. Cogn. Disord.* **2006**, *22*, 312–319. [CrossRef]
47. Ward, A.; Tardiff, S.; Dye, C.; Arrighi, H.M. Rate of conversion from prodromal Alzheimer's disease to Alzheimer's dementia: A systematic review of the literature. *Dement. Geriatr. Cogn. Dis. Extra* **2013**, *3*, 320–332. [CrossRef]
48. Salthouse, T.A. The processing-speed theory of adult age differences in cognition. *Psychol. Rev.* **1996**, *103*, 403. [CrossRef] [PubMed]
49. Crawford, T.J.; Smith, E.; Berry, D. Eye gaze and Ageing: Selective and combined effects of working memory and inhibitory control. *Front. Hum. Neurosci.* **2017**, *11*, 563. [CrossRef] [PubMed]
50. Salthouse, T.A. When does age-related cognitive decline begin? *Neurobiol. Aging* **2009**, *30*, 507–514. [CrossRef] [PubMed]
51. Peltsch, A.; Hemraj, A.; Garcia, A.; Munoz, D.P. Age-related trends in saccade characteristics among the elderly. *Neurobiol. Aging* **2011**, *32*, 669–679. [CrossRef]
52. Amatya, N.; Gong, Q.; Knox, P.C. Differing proportions of 'express saccade makers' in different human populations. *Exp. Brain Res.* **2011**, *210*, 117–129. [CrossRef]
53. Delinte, A.; Gomez, C.M.; Decostre, M.F.; Crommelinck, M.; Roucoux, A. Amplitude transition function of human express saccades. *Neurosci. Res.* **2002**, *42*, 21–34. [CrossRef]
54. Rayner, K.; Li, X.; Williams, C.C.; Cave, K.R.; Well, A.D. Eye movements during information processing tasks: Individual differences and cultural effects. *Vis. Res.* **2007**, *47*, 2714–2726. [CrossRef]
55. Alotaibi, A.; Underwood, G.; Smith, A.D. Cultural differences in attention: Eye movement evidence from a comparative visual search task. *Conscious. Cogn.* **2017**, *55*, 254–265. [CrossRef] [PubMed]
56. Chua, H.F.; Boland, J.E.; Nisbett, R.E. Cultural variation in eye movements during scene perception. *Proc. Natl. Acad. Sci. USA* **2005**, *102*, 12629–12633. [CrossRef] [PubMed]
57. Knox, P.C.; Wolohan, F.D. Cultural diversity and saccade similarities: Culture does not explain saccade latency differences between Chinese and European participants. *PLoS ONE* **2014**, *9*, e94424. [CrossRef]
58. Kim, H.S.; Sherman, D.K.; Taylor, S.E.; Sasaki, J.Y.; Chu, T.Q.; Ryu, C.; Xu, J. Culture, serotonin receptor polymorphism and locus of attention. *Soc. Cogn. Affect. Neurosci.* **2010**, *5*, 212–218. [CrossRef] [PubMed]

Article

Lacking Pace but Not Precision: Age-Related Information Processing Changes in Response to a Dynamic Attentional Control Task

Anna Torrens-Burton [1], Claire J. Hanley [2], Rodger Wood [2], Nasreen Basoudan [2], Jade Eloise Norris [3], Emma Richards [4] and Andrea Tales [4,*]

1 Division of Population Medicine, School of Medicine, Cardiff University, Cardiff CF14 4XN, UK; Torrens-BurtonA@cardiff.ac.uk
2 Department of Psychology, Swansea University, Swansea SA2 8PP, UK; c.j.Hanley@swansea.ac.uk (C.J.H.); r.l.wood@swansea.ac.uk (R.W.); nasreen.s.basoudan@gmail.com (N.B.)
3 Department of Psychology, University of Bath, Bath BA2 7AY, UK; jn528@bath.ac.uk
4 Centre for Innovative Ageing, Swansea University, Swansea SA2 8PP, UK; e.v.richards@swansea.ac.uk
* Correspondence: A.Tales@swansea.ac.uk

Received: 12 May 2020; Accepted: 14 June 2020; Published: 19 June 2020

Abstract: Age-related decline in information processing can have a substantial impact on activities such as driving. However, the assessment of these changes is often carried out using cognitive tasks that do not adequately represent the dynamic process of updating environmental stimuli. Equally, traditional tests are often static in their approach to task complexity, and do not assess difficulty within the bounds of an individual's capability. To address these limitations, we used a more ecologically valid measure, the Swansea Test of Attentional Control (STAC), in which a threshold for information processing speed is established at a given level of accuracy. We aimed to delineate how older, compared to younger, adults varied in their performance of the task, while also assessing relationships between the task outcome and gender, general cognition (MoCA), perceived memory function (MFQ), cognitive reserve (NART), and aspects of mood (PHQ-9, GAD-7). The results indicate that older adults were significantly slower than younger adults but no less precise, irrespective of gender. Age was negatively correlated with the speed of task performance. Our measure of general cognition was positively correlated with the task speed threshold but not with age per se. Perceived memory function, cognitive reserve, and mood were not related to task performance. The findings indicate that while attentional control is less efficient in older adulthood, age alone is not a defining factor in relation to accuracy. In a real-life context, general cognitive function, in conjunction with dynamic measures such as STAC, may represent a far more effective strategy for assessing the complex executive functions underlying driving ability.

Keywords: reaction time; intra-individual variability; subjective memory; healthy ageing; cognitive impairment

1. Introduction

Executive function-related reaction time (RT) and its individual variability (IIV) are regarded as behavioural markers of central nervous system integrity [1–4]. In relation to ageing, RT slowing and increased IIV are commonly described behavioural characteristics of normal, healthy processes [1,5], with disproportionate levels of decline associated with functional deficits and disease, such as mild cognitive impairment (MCI), vascular cognitive impairment (VCI), and Alzheimer's disease (AD) [6–15], as well as conditions such as mild traumatic brain injury (mTBI) [16]. It is common, therefore, for RT tests to be included in clinical diagnostic batteries.

Generally, in such tests, RT is measured over multiple sequential trials in which a response to the same given stimulus is required. Blocks of discrete trials are delivered, often with uniform task difficulty, and are analysed by focusing on each separate event or averaging responses over time [3,11,12,15]. However, the intricacies of visual perception in the real-world are far more complex than that represented by the measurement of RT in this way. For example, mediated by attentional control processes [17], representations are constantly updated to inform the individual of the crucial elements in scenes of varying complexity [18,19], in driving for example [16,20]. This means that approaches that adopt discrete trials cannot mimic the manner in which attentional control operates, nor can they represent RT in its real-life context, thus limiting the validity of the inferences that can be made when utilizing such methods.

1.1. Attentional Control

Attentional control processes are associated with a network involving prefrontal and parietal regions, which govern target detection and response inhibition, under conditions of varying difficulty and demand for resources [21,22], all of which in real-life would contribute to a measure of RT at any given point. Although there is some evidence to suggest that for lower task demands older adults can maintain relatively intact performance (as a result of an ability to increase recruitment of available neural resources to stabilise performance), age-related decreases in activity tend to occur with higher task demands [23–25] because available processing resources are saturated [17]. Consequently, age-related failure to modulate activity in response to changes in task demand may indicate an inability to dynamically allocate attentional control in response to change. Furthermore, although difficulty can be varied in common attention-related RT tasks such as those examining visual search, inhibition of return, alerting, and orienting [14,26–30], it is not typically investigated with respect to individual capability (e.g., accommodating differences in what individual participants find challenging), which is an important consideration when devising tasks for use with a highly heterogeneous population, such as older adults.

Typical RT studies also simply examine the relationship between error production and RT, in order to determine whether a speed/accuracy trade-off is influencing the results [26,28,30,31]. In reality, however, the interplay between speed and accuracy is likely to be far more relevant to real life environmental processing. When responding appropriately to environmental change, a certain level of accuracy is required as well as speed. It is likely that the ability to maintain accuracy, whilst also maintaining speed, is a more adaptive behaviour than the ability to respond quickly to a given stimulus that is repeatedly presented, with no change in resource requirements. Examining RT in isolation from the integrity of attentional control, and thus the above-mentioned functions, may lead to the under- or over- estimation of an individual's functional ability; a factor which may be of particular relevance to the assessment of driving ability [32] as this behaviour is highly dependent upon the integrity of attentional control components such as selective attention, attentional switching, and the inhibition of irrelevant visual information [33–37]. As driving cessation is widely considered to be a major life transition which can have a significant impact on the health and well-being of older drivers [38–40], it is vital that such ability is assessed appropriately [32,38,41].

1.2. Study Rationale

In this study, we examined ageing-related attentional control (i.e., how much information can be processed over a specific time period, at a given level of accuracy) using a novel task, designed to address the outlined issues with traditional tests; namely the computer-based Swansea Test of Attentional Control (STAC). The STAC (developed by Carter and Wood, and outlined in Hanley and Tales, (2019) [42] comprises selective attention, task monitoring, and response inhibition components of attentional control based on the supervisory attentional system model [22], making it ideal to simulate the complex demands of continuous environmental monitoring and interaction, within a single test. A flexible algorithm designed to track performance (Parameter Estimation by Sequential Testing,

PEST; [43]), calibrates the speed at which a given degree of response accuracy can be maintained. STAC also relies on the continuous presentation of stimuli, as opposed to delivering discrete trials in blocks that systematically vary task demands [17,44]. Therefore, compared to other attentional control tasks in isolation, STAC has the benefit of being more holistic in relation to the demands of stimulus engagement in everyday life. Furthermore, the stability and validity of the STAC task have previously been assessed, with good test–retest reliability and strong correspondence between STAC final speed and RTs from a standard Flanker task (for further insight and information about this test, see Hanley and Tales, (2019) [42]). Unlike traditional tests, which focus on single elements of attentional control in isolation, STAC is more holistic and integrates components of selective attention, task monitoring, and response inhibition. Furthermore, while there is overlap in the constructs tested, standard tasks such as the Flanker test (known for its ability to assess response inhibition) require the researcher to set parameters (e.g., stimulus presentation speed) in advance, and often lack flexibility because such values are fixed, meaning the participant is forced to struggle to respond or is not sufficiently challenged. Use of the PEST algorithm as part of STAC, which calibrates speed on the basis of prior responses, has the distinct advantage of ensuring that the task is performed in accordance with an individual's capabilities. Therefore, the task remains difficult and will challenge the participant as speed is re-adjusted during subsequent PEST cycles (either upwards or downwards, to define their threshold, in the event that the present speed can be exceeded or is too challenging and cannot be sustained).

The comparison of STAC to the Flanker task corresponds to our previous pilot work. While the Flanker task output equates to reaction time (RT), the STAC final speed threshold reflects a participants' ability to perform well under conditions of higher stimulus presentation speeds. Nonetheless, both measures represent efficiency of processing with regard to speed. Where RT from incongruent Flanker trials (representing conditions of maximum difficulty) is assessed in relation to STAC speed, there is strong correspondence between these measures (r (24) = −0.650, $p = 0.001$). The correlation indicates that as STAC speed increases, Flanker RT decreases, thus signifying performance improvement between the measures is aligned. Therefore, STAC is regarded as comparable to Flanker in relation to outcome measures but has distinct advantages compared to such standard tasks, as outlined above.

Adopting this novel task, the primary aim of the study was to examine attentional control in older compared to younger adults. Within the scope of test complexity and task difficulty, whilst task difficulty is not a factor under direct manipulation, with discrete levels of complexity, it corresponds to the speed of performance (spm); whereby performing the task at a higher speed would increase the demand for processing resources. Accordingly, higher speed thresholds represent the ability of participants to successfully complete the task under conditions of increased difficulty.

We also aimed to determine whether attentional control varied with respect to general cognitive function (using the Montreal Cognitive Assessment (MoCA)) [45], typically used to define an older adult control group. For the older adults, we also examined performance with respect to subjective feelings of memory function (using the Memory Function Questionnaire (MFQ) [46], cognitive reserve (using the National Adult Reading Test, NART score as a proxy) and educational level [47–52]. Finally, a mix of male and female older adult participants was included, all with depression and anxiety within normal levels.

In relation to these aims, we predicted that attentional control would be significantly poorer in older compared to younger adults, as determined by speed of performance on the STAC task; and that high levels of cognitive control would be significantly associated with high cognitive function, high cognitive reserve, and lower levels of perceived abnormal memory. We also investigated whether attentional control varied with respect to gender and sub-clinical levels of depression and anxiety [13,53,54].

By obtaining this range of measurements we were well positioned to determine how, if at all, attentional control is related to the level of objectively and subjectively measured cognitive function, within individuals. This is a crucial consideration as older adults with ostensibly normal cognition for their age and educational level, and some individuals with subjective cognitive decline (SCD) [55], can show considerable variation in RT and attention. Some individuals may even present with levels of detrimental functional change more typical of MCI, vascular cognitive impairment (VCI), and Alzheimer's disease (AD) [11–14].

2. Materials and Methods

2.1. Participants

Community-dwelling older adults (n = 90; age 50–79 years; 38 males, 52 females) were recruited through advertisements at older adult social clubs, via local newspapers, word of mouth, and via the older adult research volunteer database (Department of Psychology, Swansea University), with no specified upper age limit. Younger adults, aged between 18 and 30 years and thus typical of younger control groups in ageing studies, (n = 82; age 18–27 years; 32 males, 50 females) were recruited via poster, social networking, and email advertisements throughout Swansea University. Testing was conducted within dedicated research rooms within the Department of Psychology at Swansea University. Ethical approval was granted by the Swansea University Department of Psychology Research Ethics Committee (Reference number: 01-20/2-2015-1), and the study was conducted in accordance with the principles of the Declaration of Helsinki. Written, informed consent was given by each participant, and all participants were debriefed after participation.

All participants self-reported to be in good general health, with no history of serious head injury, mental health problems, or neurological impairment. None of the participants had been to see their general practitioner (GP) or memory services about change in their cognitive function or concern about it, and all participants had normal or corrected-to-normal vision. The groups were matched with respect to gender distribution and educational level (Table 1). None of the participants had significant levels of depression or anxiety, as indicated by a score of 9 or below on the Patient Health Questionnaire (PHQ-9) [56] and a score of 5 or below on the Generalized Anxiety Disorder 7-item (GAD-7) [57] (using pen and paper-based versions of both of these tests) and individuals with a history of significant depression or anxiety were excluded from this study. Although medication use could not be controlled, individuals taking drugs likely to affect cognition (such as benzodiazepines) were excluded from recruitment.

Table 1. Baseline group mean demographics for the younger and older adult groups. Standard deviation in parenthesis.

	Age (Years)	Education Level (Years)	NART	MoCA Score	MFQ (Total Score)	PHQ-9	GAD-7
Young Adults	20 (2.1)	14.7 (3.5)	-	27.0 (2.1)	-	6.3 (4.4)	4.8 (4.3)
Older Adults	65 (6.1) Range 50–79	15.1 (4.8)	41.4 (5.4)	27.4 (2.2)	292.8 (48.8)	3.2 (3.1)	2.3 (2.5)

All data were analysed using SPSS version 22 (IBM Corp., Armonk, NY, USA).

2.2. Materials and Procedures

Cognitive function was assessed using the pen and paper-based Montreal Cognitive Assessment (MoCA) [45]. For the older adults, subjective memory function was assessed by the pen and paper-based Memory Functioning Questionnaire (MFQ) [46] (in which lower scores represent a greater degree of perceived memory impairment). The NART was administered to the older adult participants as a proxy for cognitive reserve [47–49]. The order of these tests was not counterbalanced, and all tests were administered and scored according to current guidelines. The computer-based STAC test was then administered.

Symbols are 36.2 mm across, subtending 3.6 degrees of visual angle on a 19-inch monitor at 52 cm viewing distance. A target is identified within the 3 × 3 matrix of symbols (right). When a matching symbol appears amongst the three columns of the search array (left), which scroll up the left-hand side of the screen, participants press the spacebar as the symbol crosses behind the red line. The task is to identify the target within a 3 × 3 matrix of symbols on the right and search for matching symbols amongst an array of three columns of symbols (Figure 1). The target changes throughout the task such that participants must remain vigilant in order to consistently update their search criteria, while simultaneously monitoring the search array to identify matching items and inhibit irrelevant symbols. The target changes every 19 s but is delayed if the current target appears in the search array (e.g., the target will not change to a different symbol if the current target is on screen). In such instances, the corresponding time lapse is added to the total run time (initially set to 180 s, resulting in ~9 target changes for all participants). Speed (measured in symbols per minute per column; abbreviated to "spm") is adjusted to maintain accuracy around a commonly adopted 75% correct criterion [58], using the PEST algorithm. The task begins at a speed of 60 spm. After a minimum of 4 target changes, speed is calibrated on the basis of performance accuracy, increasing or decreasing with a step-size between 20 and 60 spm. The participants' thresholds are the average speed at which the task is performed across the duration of the task.

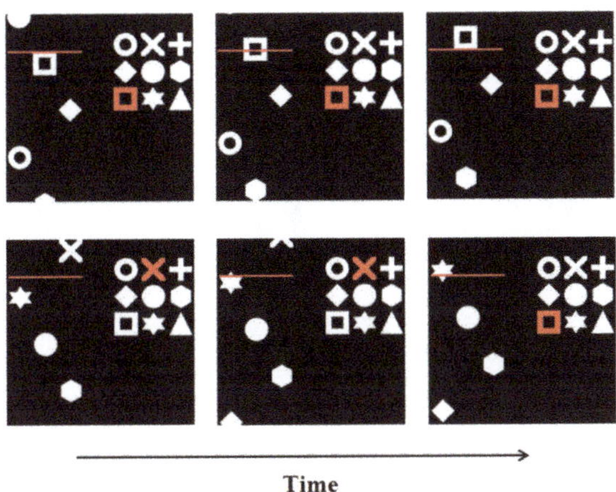

Figure 1. Swansea Test of Attentional Control. Participants view a 3 × 3 matrix of symbols (screen right) and a continuous stream of symbols (screen left). Top: The symbols on the left scroll up the screen. Where an oncoming symbol matches the target (highlighted in red, on the right), participants respond when that shape crosses the red line (as opposed to before or after). Bottom: The target changes every 19 s, such that participants must remain vigilant and consistently attend to both elements of the task to ensure successful performance.

2.3. Data Analysis and Results

With respect to demographics, Mann–Whitney analysis revealed no significant difference in MoCA score or mean years in education between young and older adult groups ($p > 0.05$). Depression level (U = 1911.5, $p < 0.001$, $r = 0.37$) and anxiety level (U = 1638, $p < 0.001$, $r = 0.32$) were significantly greater for young adults compared to older adults.

Non-parametric analyses were performed as the STAC data were not normally distributed.

Mann–Whitney U analysis revealed no significant difference in mean accuracy between the older and younger adults (U = 3580.5, $p = 0.74$) and for both the younger and older groups this did not vary significantly with respect to gender ((male U = 785.5, $p = 0.89$) and (female U = 946.5, $p = 0.73$)). There was, however, a significant difference in spm, with younger adults able to process significantly more symbols per minute per column than the older adults, performing the task significantly faster (U = 2059.5, $p < 0.001$; Cohen's effect size = 0.884) (Table 2 and Figure 2). There was no significant difference in spm between male and female participants, for both the younger (U = 602.5, $p = 0.06$) and older groups (U = 978.5, $p = 0.94$), respectively.

Table 2. Symbols per minute per column (spm) and level of accuracy (%) for younger and older adults. Standard deviation in parentheses.

	Mean spm	% Accuracy
Young Adults	94.78 (29.69)	83.82 (9.15)
Older Adults	71.74 (22.26)	83.16 (13.63)

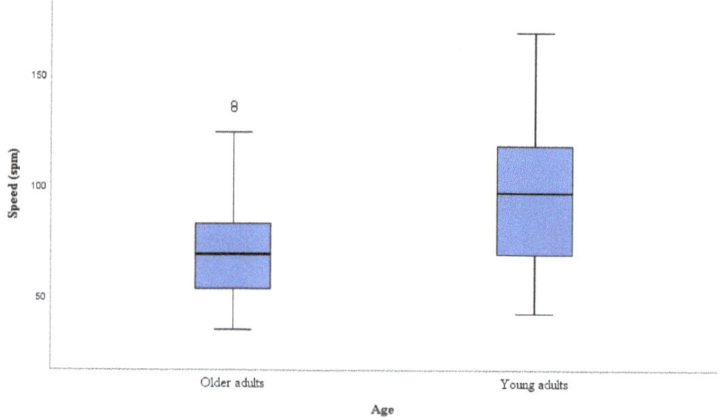

Figure 2. Speed (symbols per minute per column; spm) between young and older adults.

For the older adults, Spearman's rho analyses revealed a significant correlation between spm and MoCA score ($r = 0.25$, $p = 0.018$), with better performance associated with higher MoCA scores (higher levels of general cognitive function). In contrast, there was no significant correlation between spm and MoCA score ($p > 0.05$) for the younger adults.

There was also a significant negative correlation between spm and age ($r = -0.29$, $p = 0.005$) (as age increased, spm decreased) for the older adults (Figure 3). However, there was no significant correlation between age and MoCA score ($p > 0.05$). For the younger adults, there was no significant correlation between spm and age ($p > 0.05$), and no significant correlation between age and MoCA score ($p > 0.05$). For the older adults, spm was not significantly correlated with MFQ or NART, and there was no significant correlation between MFQ and MoCA scores (all p-values > 0.05). For both the younger and older adults, spm was not significantly correlated with anxiety, depression, or educational level (all p-values > 0.05).

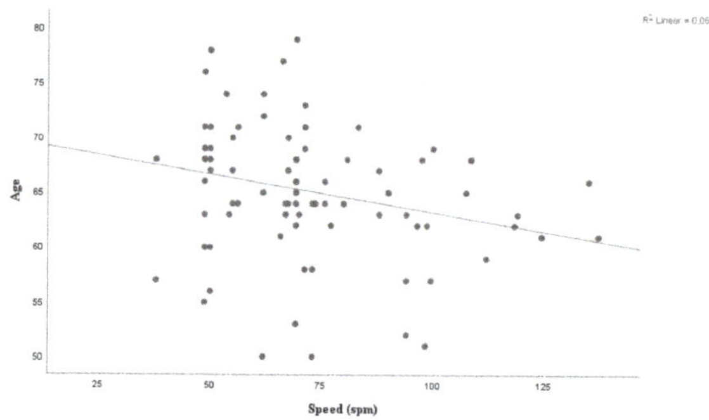

Figure 3. Correlation between speed (spm) and MoCA score in older adults.

3. Discussion

In this study, we used the STAC test to examine the complex interplay between the speed and accuracy of information processing in older compared to younger adults. As predicted, the requirement to maintain a stimulus accuracy response of at least 75% (actually achieving approximately 84% in this study) resulted in a reduction in the amount of processing possible within a given time period. Specifically, this was characterised by a significant decrease in the speed at which the symbols per minute per column could be processed, for the older compared to the younger group. The ability to modulate activity to varying task resource demands and inhibit interference while maintaining a high level of processing speed for a given accuracy of performance (attentional control), therefore, appears significantly less efficient in older adulthood. This is likely a result of a reduced ability to dynamically allocate attention, such that supply is equal to demand. These results, which were independent of gender, are in line with well-established evidence showing that higher demand for resources, under conditions of greater complexity, causes issues in relation to the efficacy of information processing with age [17,23–25]. Unlike previous studies, we employed methodology that allowed for within-group variation in respect to individual capability (i.e., accommodating differences in what participants find challenging), thus removing a potential confounding factor present in previous studies with variation in task difficulty; an important consideration when devising tasks for use with a highly heterogeneous older adult population.

We were also interested in examining any relationship between general cognition measured using the MoCA [45] in individuals who had not reported any cognitive changes or concerns to their general practitioner/memory services, and STAC performance. For the older adults there was a significant correlation between spm and MoCA score, with higher levels of general cognition function associated with better performance; a significant negative correlation between spm and age with spm decreasing as age increased, but there was no significant correlation between age and MoCA score. This pattern of results suggests that the reduction in speed of performance as age increases may be compensated for (at least within the age range of the participants in this study), by higher levels of objectively measured levels of cognitive function. The lack of a significant correlation between spm and MFQ score indicates that attentional control is not however associated with, or compensated for, by perceived level of cognitive function.

Attentional control in older adults was also not a function of cognitive reserve, (using the proxy measure of NART score), although one could argue that NART score was not representative of the complexity of cognitive reserve, nor of all its components. In contrast to the older adults, there was no significant correlation between spm and age, or spm and MoCA score, for the younger adults. For both the younger and older adults there was no significant correlation between spm and anxiety, depression or educational level, indicating that attentional control was not associated with such factors at the levels or ranges encountered in the present study.

With respect to older adults, the significant correlation between spm and MoCA score, significant negative correlation between spm and age, and the lack of significant correlation between age and MoCA score, indicates that chronological age alone is unlikely to be useful in determining real life behaviour such as driving. This supports the idea that functional ageing, as opposed to chronological ageing, may be a more appropriate indicator of real-life ability. The results indicate that although one gets slower with increasing age, accuracy is still maintained. This is likely true up to a point, or certain age; after which it may no longer be appropriate for older adults to engage in driving behaviour that is more "resource intensive" (such as driving on motorways or at night, which requires higher speeds and increased vigilance). Although the correlation between MoCA score and STAC performance (spm) may provide some insight into how well an individual can manage these demands, it may be more appropriate if lower scores were used to prompt more in-depth cognitive assessment, such as use of STAC, to assess driving behaviour [16,20,32,38,41], particularly as driving cessation is widely considered to be a major life transition, and can have a significant impact on the health and well-being of older drivers [39,40].

3.1. Potential Study Limitations

We have previously shown that older adults (mean age 66.5 years) are able to comfortably perform the STAC at a length of 300 s [42]. However, due to the novel nature of the task, the "optimal" duration for both younger and older adults is yet to be determined. In this instance, the task was only run for a short amount of time (180 s), which may have favoured performance in the younger group more so than the older group, because for the latter it has been found to improve with continued exposure. Therefore, the distinction between groups may be diminished at longer intervals. Furthermore, although the PEST algorithm should naturally converge on a participant's optimal speed, given sufficient time, it would certainly be an intriguing prospect to assess performance capability where the start speed of the experiment was founded on the basis of an individual's threshold (established during practice, for example).

3.2. Future Studies

Without accompanying neuroimaging data, the study is not able to make direct inferences relating to the efficiency of neural processes supporting attentional control. In future, it would be of interest to use task-based fMRI to assess the availability of processing resources in given regions or networks (prefrontal/parietal). Use of diffusion-weighted imaging to assess the integrity of key tracts related to attention (genu/forceps minor, superior longitudinal fasciculus) could also be incorporated. Additional electroencephalographic (EEG) analysis both in the time-domain and in the time-frequency domain would also reveal the recruitment of additional cognitive resources required for the maintenance of the accuracy with increasing age. The examination of the variations in task difficulty and complexity based on the individual threshold of a single participant (e.g., a variation of a specific % of spm), would also provide a unique insight into the neural underpinnings of this novel task, age-related distinctions, and individual differences.

The STAC represents a complex task that requires good attention control to optimise efficient information processing. These processes rely on sub-cortical white matter integrity for speed and efficiency of processing, and the pre-frontal cortex to "supervise" the allocation of attentional resources. These cerebral regions are vulnerable to the decelerative forces present in traumatic brain injury and often underpin weaknesses of cognition reported by patients, even after relative minor head trauma. However, identifying attentional failures during a clinical assessment can be difficult [58]. This probably accounts for the dislocation, often seen when a patient performs well on clinical tests of cognition yet complains of intrusive lapses of working memory and other cognitive failures in everyday life. This disparity is assumed to reflect the different attentional loadings present in a clinical assessment versus everyday life. As the STAC replicates the changing attentional demands encountered in the cognitive control of everyday activities, it may also be an important tool for the investigation of head injury with respect to the determination functional integrity and response to intervention.

Furthermore, cognitive function was only assessed using the MoCA [45] and thus only a composite score of performance over several cognitive domains, namely short-term memory, visuospatial ability, executive function, attention, concentration and working memory, language, and orientation to time and place. Future studies will involve the examination of a wider range of specific cognitive functions in order to examine the links between cognitive function and attentional control, information processing speed, and accuracy.

Finally, it would be of value to administer the task for a range of durations to assess age-related differences under conditions of varying demands on sustained attention. Furthermore, the age-range was relatively large in the present study, and although there was no specified upper age limit our older adult group contained no-one over the age of 79 years and future studies should focus on a smaller age range or sub-divide the older age group into different cohorts (e.g., 50–65, 66–80) and to include individuals above the age of 80 in order to assess age effects in more detail.

Author Contributions: A.T., C.J.H. and R.W. conceived and designed the experiments; A.T.-B., N.B., E.R., performed the experiments; A.T., C.J.H., A.T.-B., and J.E.N., analyzed the data; A.T., A.T.-B., C.J.H., R.W., N.B., J.E.N., and E.R., wrote the paper. All authors have read and agreed to the published version of the manuscript.

Funding: This Research was funded by BRACE- Dementia Research. Registered Charity: 297965. The APC was funded by Swansea University.

Acknowledgments: The authors would like to acknowledge and thank BRACE- Dementia Research. Registered Charity: 297965 for funding this research, Neil Carter for help in developing the STAC test and to all the people who very kindly gave up their time to take part in this study.

Conflicts of Interest: The authors declare no conflict of interest.

References

1. Kuznetsova, K.A.; Maniega, S.M.; Ritchie, S.J.; Cox, S.R.; Storkey, A.J.; Starr, J.M.; Wardlau, J.M.; Deary, I.J.; Bastin, M.E. Brain white matter structure and information processing speed in healthy older age. *Brain. Struct. Funct.* **2016**, *221*, 3223–3235. [CrossRef]
2. Yang, Y.; Bender, A.R.; Raz, N. Age related differences in reaction time components and diffusion properties of normal-appearing white matter in healthy adults. *Neuropsychologia* **2015**, *66*, 246–258. [CrossRef] [PubMed]
3. Haynes, B.I.; Bauermeister, S.; Bunce, D. A systematic review of longitudinal associations between reaction time intraindividual variability and age-related cognitive decline or impairment, dementia and mortality. *J. Int. Neuropsychol. Soc.* **2017**, *23*, 431–445. [CrossRef] [PubMed]
4. Hong, Z.; Ng, K.K.; Sim, S.K.; Ngeow, M.Y.; Zheng, H.; Lo, J.C.; Chee, M.W.; Zhou, J. Differential age-dependent associations of gray matter volume and white matter integrity with processing speed in healthy older adults. *Neuroimage* **2015**, *123*, 42–50. [CrossRef] [PubMed]
5. Bielak, A.A.M.; Cherbuin, N.; Bunce, D.; Anstey, K.J. Intraindividual variability is a fundamental phenomenon of Aging: Evidence from an 8-year longitudinal study across young, middle and older adulthood. *Develop. Psych.* **2014**, *50*, 143–151. [CrossRef]
6. Jackson, J.D.; Balota, D.A.; Duchek, J.M.; Head, D. White matter integrity and reaction time intraindividual variability in healthy aging and early-stage Alzheimer's Disease. *Neuropsychologia* **2012**, *50*, 357–366. [CrossRef]
7. Bunce, D.; Haynes, B.I.; Lord, S.R.; Gschwind, Y.J.; Kochan, N.A.; Reppermund, S.; Brodaty, H.; Sachdev, P.S.; Delbaere, K. Intraindividual stepping reaction time variability predicts falls in older adults with mild cognitive impairment. *J. Gerontol. A Biol. Sci. Med. Sci.* **2016**, *72*, 832–837. [CrossRef]
8. Aichele, S.; Rabbitt, P.; Ghisletta, P. Think fast, feel fine, live long: A 29-year study of cognition, health, and survival in middle-aged and older adults. *Psychol. Sci.* **2016**, *4*, 518–529. [CrossRef]
9. Kochan, N.A.; Bunce, D.; Pont, S.; Crawford, J.D.; Brodaty, H.; Sachdev, P.S. Is intraindividual reaction time variability an independent cognitive predictor of mortality in old age? Findings from the Sydney memory and ageing study. *PLoS ONE* **2017**, *12*, e0181719. [CrossRef]
10. Troyer, A.K.; Vandermorris, S.; Murphy, K.J. Intraindividual variability on associative memory tasks is elevated in amnestic mild cognitive impairment. *Neuropsychologia* **2016**, *90*, 110–116. [CrossRef]
11. Richards, E.; Bayer, A.; Hanley, C.; Norris, J.E.; Tree, J.J.; Tales, A. Reaction time and visible white matter lesions in subcortical ischemic vascular cognitive impairment. *J. Alz. Dis.* **2019**, *72*, 859–865. [CrossRef] [PubMed]
12. Richards, E.; Bayer, A.; Hanley, C.; Norris, J.E.; Tales, A. Subcortical ischemic vascular cognitive impairment: Insights from reaction time measures. *J. Alz. Dis.* **2019**, *723*, 845–857. [CrossRef]
13. Tales, A.; Leonards, U.; Bompas, A.; Snowden, R.; Phillips, M.; Porter, G.; Hawworth, J.; Wilcock, G.; Bayer, A. Intra-individual reaction time variability in amnestic mild cognitive impairment: A precursor to dementia? *J. Alz. Dis.* **2012**, *32*, 457–466. [CrossRef]
14. Bayer, A.; Phillips, M.; Porter, G.; Leonards, U.; Bompas, A.; Tales, A. Abnormal inhibition of return in mild cognitive impairment: Is it specific to the presence of prodromal dementia? *J. Alz. Dis.* **2014**, *40*, 177–189. [CrossRef] [PubMed]
15. Haworth, J.; Phillips, M.; Newson, M.; Rogers, P.J.; Torrens-Burton, A.; Tales, A. Measuring information processing speed in mild cognitive impairment: Clinical versus research dichotomy. *J. Alz. Dis.* **2016**, *51*, 263–275. [CrossRef] [PubMed]

16. Panwar, N.; Purohit, D.; Deo Sinha, V.; Joshi, M. Evaluation of extent and pattern of neurocognitive functions in mild and moderate traumatic brain injury patients by using Montreal Cognitive Assessment (MoCA) score as a screening tool: An observational study from India. *Asian J. Psychiatry* **2019**, *41*, 60–65. [CrossRef]
17. Hedden, T.; Van Dijk, K.R.A.; Shire, E.H.; Sperling, R.A.; Johnson, K.A.; Buckner, R.L. Failure to modulate attentional control in advanced aging linked to white matter pathology. *Cereb. Cortex.* **2012**, *22*, 1038–1051. [CrossRef]
18. Rensink, R.A. The dynamic representation of scenes. *Vis. Cogn.* **2000**, *7*, 17–42. [CrossRef]
19. Guest, D.; Howard, C.J.; Brown, L.A.; Gleeson, H. Aging and the rate of visual information processing. *J. Vis.* **2015**, *15*, 1–25. [CrossRef]
20. Dymowski, A.R.; Owens, J.A.; Ponsford, J.L.; Willmott, C. Speed of processing and strategy control of attention after traumatic brain injury. *J. Clin. Exp. Neuropsychol.* **2015**, *37*, 1024–1035. [CrossRef]
21. Sylvester, C.Y.C.; Wager, T.D.; Lacey, S.C.; Hernandez, L. Switching attention and resolving interference: fMRI measures of executive functions. *Neuropsychologia* **2003**, *41*, 357–370. [CrossRef]
22. Cieslik, E.C.; Mueller, V.I.; Eickhoff, C.R.; Langner, R.; Eickhoff, S.B. Three key regions for supervisory attentional control: Evidence from neuroimaging meta-analyses. *Neurosci. Biobehav. Rev.* **2015**, *48*, 22–34. [CrossRef] [PubMed]
23. Maylor, E.A.; Wing, A.M. Age differences in postural stability are increased by additional cognitive demands. *J. Gerontol. Ser. B Psychol. Sci. Soc. Sci.* **1996**, *51*, 143–154. [CrossRef] [PubMed]
24. Teasdale, N.; Bard, C.; LaRue, J.; Fleury, M. On the cognitive penetrability of posture control. *Exp. Aging Res.* **1993**, *19*, 1–3. [CrossRef] [PubMed]
25. Maylor, E.A.; Lavie, N. The influence of perceptual load on age differences in selective attention. *Psychol. Aging* **1998**, *13*, 563–573. [CrossRef]
26. Tales, A.; Muir, J.; Jones, R.; Bayer, A.; Snowden, R.J. The effects of saliency and task difficulty on visual search performance in ageing and Alzheimer's disease. *Neuropsychologia* **2004**, *42*, 335–345. [CrossRef] [PubMed]
27. Vaportzis, E.; Georgiou-Karistianis, N.; Stout, J.C. Dual task performance in normal aging: A compaison of choice reaction time tasks. *PLoS ONE* **2013**, *8*, e60265. [CrossRef]
28. Tales, A.; Snowden, R.J.; Phillips, M.; Haworth, J.; Porter, G.; Wilcock, G.; Bayer, A. Exogenous phasic alerting and spatial orienting in mild cognitive impairment compared to healthy ageing: Study outcome is related to target response. *Cortex* **2011**, *47*, 180–190. [CrossRef]
29. Eimer, M. The neural basis of attentional control in visual search. *Trends Cogn. Sci.* **2014**, *18*. [CrossRef]
30. Torrens-Burton, A.; Basoudan, N.; Bayer, A.; Tales, A. Perception and reality of cognitive function: Information processing speed, perceived memory function and perceived task difficulty in older adults. *J. Alz. Dis.* **2017**, *60*, 1601–1609. [CrossRef]
31. Heitz, R.P. The speed-accuracy tradeoff: History, physiology, methodology, and behaviour. *Front. Neurosci.* **2014**, *8*, 150. [CrossRef] [PubMed]
32. Hird, M.A.; Egeto, P.; Fischer, C.E.; Naglie, G.; Schweizer, T.A. A systematic review and meta-analysis of on-road simulator and cognitive driving assessment in Alzheimer's disease and mild cognitive impairment. *J. Alz. Dis.* **2016**, *53*, 713–729. [CrossRef] [PubMed]
33. Callaghan, E.; Holland, C.; Kessler, K. Age-related changes in the ability to switch between temporal and spatial attention. *Front. Ageing Neurosci.* **2017**, *9*, 28. [CrossRef]
34. Stinchcombe, A.; Gagnon, S.; Zhang, J.J.; Montembeault, P.; Bedard, M. Fluctuating attentional demand in the simulated driving assessment: The roles of age and driving complexity. *Traffic Inj. Prev.* **2011**, *12*, 576–587. [CrossRef] [PubMed]
35. Tsang, P.S. Ageing and attentional control. *Q. J. Exp. Psychol.* **2013**, *66*, 1517–1547. [CrossRef] [PubMed]
36. Roca, J.; Castro, C.; López-Ramón, M.F.; Lupiáñez, J. Measuring vigilence while assessing the function of the three attentional networks: The ANI-Vigilance task. *J. Neurosci. Methods.* **2011**, *198*, 312–324. [CrossRef]
37. McManus, B.; Heaton, K.; Stavrinos, D. Commercial motor vehicle driving performance: An examination of attentional resources and control using a driving simulator. *J. Exp. Psychol. Appl.* **2017**, *23*, 191–203. [CrossRef]
38. Pyun, J.-M.; Kang, M.J.; Kim, S.; Baek, M.J.; Wang, M.J.; Kim, S.Y. Driving cessation and cognitive dysfunction in patients with mild cognitive impairment. *J. Clin. Med.* **2018**, *7*, 545. [CrossRef]
39. Liddle, J.; Turpin, M.; Carlson, G.; McKenna, K. The needs and experiences related to driving cessation for older people. *Br. J. Occup. Therapy* **2008**, *71*, 379–388. [CrossRef]

40. Musselwhite, C.B.; Shergold, I. Examining the process of driving cessation in later life. *EU. J. Ageing* **2013**, *10*, 89–100. [CrossRef]
41. Whelihan, W.M.; DiCarlo, M.A.; Paul, R.H. The relationship of neuropsychological functioning to driving competence in older persons with early cognitive decline. *Arch. Clin. Neuropsychol.* **2005**, *20*, 217–228. [CrossRef] [PubMed]
42. Hanley, C.J.; Tales, A. Anodal tDCS improves attentional control in older adults. *Exp. Gerontol.* **2019**, *115*, 88–95. [CrossRef] [PubMed]
43. Taylor, M.; Creelman, C.D. PEST: Efficient estimates on probability functions. *J. Acoust. Soc. Am.* **1967**, *41*, 782–787. [CrossRef]
44. Aschenbrenner, A.J.; Balota, D.A. Dynamic adjustments of attentional control in healthy aging. *Psychol. Aging* **2017**, *32*, 1–15. [CrossRef] [PubMed]
45. Nasreddine, Z.S.; Phillips, N.A.; Bédirian, V.; Charbonneau, S.; Whitehead, V.; Collin, I.; Cummings, J.L.; Chertkow, H. The Montreal Cognitive Assessment, MoCA: A brief screening tool for mild cognitive impairment. *J. Am. Geriatr.* **2005**, *53*, 695–699. [CrossRef]
46. Gilewski, M.J.; Zelinski, E.M.; Schaie, K.W. The Memory functioning questionnaire for assessment of memory complaints in adulthood and old age. *Psychol. Aging* **1990**, *5*, 482–490. [CrossRef]
47. Nelson, H.E.; Willison, J.R. *National Adult Reading Test (NART). Test Manual Including New Data Supplement, NFER-Nelson, Windsor*; NFER-Nelson Publishing Company Ltd.: London, UK, 1991.
48. Stern, Y. Cognitive Reserve. *Neuropsychologia* **2009**, *47*, 2015–2028. [CrossRef]
49. Dykiert, D.; Deary, I.J. Retrospective validation of WTAR and NART scores as estimators of prior cognitive ability using the Lothian Birth Cohort 1936. *Psychol. Assess.* **2013**, *25*, 1361–1366. [CrossRef]
50. Olaithe, M.; Bucks, R.S.; Eastwood, P.; Hillman, D.; Skinner, T.; James, A.; Hunter, M.; Gavett, B. A brief report: The national adult reading test (NART) is a stable assessment of premorbid intelligence across disease severity in obstructive sleep apnea (OSA). *J. Sleep Res.* **2019**. [CrossRef]
51. Osone, A.; Arai, R.; Hakamada, R.; Shimoda, K. Cognitive and brain reserve in conversion and reversion in patients with mild cognitive impairment over 12 months of follow-up. *J. Clin. Exp. Neuropsychol.* **2016**, *38*, 1084–1093. [CrossRef]
52. Alosco, M.L.; Spitznagel, M.B.; Raz, N.; Cohen, R.; Sweet, L.H.; Van Dulmen, M.; Colbert, L.H.; Josephson, R.; Waechter, D.; Hughes, J.; et al. Cognitive reserve moderates the association between heart failure and cognitive impairment. *J. Clin. Exp. Neuropsychol.* **2012**, *34*, 1–10. [CrossRef] [PubMed]
53. Tales, A.; Basoudan, N. Anxiety in old age and dementia- implications for clinical and research practice. *Neuropsychiatry* **2016**, *6*, 142–148. [CrossRef]
54. Reimers, S.; Maylor, S. Gender effects on reaction time variability and trial-to-trial performance: Reply to Deary and Der. *Aging Neuropsychol. Cogn.* **2005**, *13*, 479–489. [CrossRef] [PubMed]
55. Jessen, F.; Amariglio, R.E.; Buckley, R.F.; Van Der Flier, W.M.; Han, Y.; Molinuevo, J.L.; Rabin, L.; Rentz, D.M.; Rodrigo-Gomez, O.; Saykin, A.J.; et al. The characterization of subjective cognitive decline. *Lancet Neurol.* **2020**, *19*, 271–278. [CrossRef]
56. Kroenke, K.; Spitzer, R.L.; Williams, J.B. The PHQ-9. *J. Gen. Intern. Med.* **2001**, *16*, 606–613. [CrossRef]
57. Löwe, B.; Decker, O.; Müller, S.; Brähler, E.; Schellberg, D.; Herzog, W.; Herzberg, P.Y. Validation and standardization of the Generalized Anxiety Disorder Screener (GAD-7) in the general population. *Med. Care.* **2008**, *46*, 266–274. [CrossRef]
58. Wood, R.L.l.; Bigler, E. Problems Assessing Executive Dysfunction in Neurobehavioural Disability. In *Neurobehavioural Disability and Social Handicap Following Traumatic Brain Injury*; McMillan, T.M., Wood, R.L., Eds.; Psychology Press and Routledge Classic: Oxfordshire, UK, 2017.

Article

Age-Related Changes in Attentional Refocusing during Simulated Driving

Eleanor Huizeling [1,2,*,†], Hongfang Wang [1], Carol Holland [2,‡] and Klaus Kessler [1,2,*]

1 Aston Neuroscience Institute, Aston University, Birmingham B4 7ET, UK; h.wang26@aston.ac.uk
2 Aston Research Centre for Healthy Ageing, Aston University, Birmingham B4 7ET, UK; c.a.holland@lancaster.ac.uk
* Correspondence: Eleanor.Huizeling@mpi.nl (E.H.); k.kessler@aston.ac.uk (K.K.)
† Present address: Max Planck Institute for Psycholinguistics, 6525 XD Nijmegen, The Netherlands.
‡ Present address: Centre for Ageing Research, Division of Health Research, Lancaster University, Lancaster LA1 4YW, UK.

Received: 14 July 2020; Accepted: 5 August 2020; Published: 7 August 2020

Abstract: We recently reported that refocusing attention between temporal and spatial tasks becomes more difficult with increasing age, which could impair daily activities such as driving (Callaghan et al., 2017). Here, we investigated the extent to which difficulties in refocusing attention extend to naturalistic settings such as simulated driving. A total of 118 participants in five age groups (18–30; 40–49; 50–59; 60–69; 70–91 years) were compared during continuous simulated driving, where they repeatedly switched from braking due to traffic ahead (a spatially focal yet temporally complex task) to reading a motorway road sign (a spatially more distributed task). Sequential-Task (switching) performance was compared to Single-Task performance (road sign only) to calculate age-related switch-costs. Electroencephalography was recorded in 34 participants (17 in the 18–30 and 17 in the 60+ years groups) to explore age-related changes in the neural oscillatory signatures of refocusing attention while driving. We indeed observed age-related impairments in attentional refocusing, evidenced by increased switch-costs in response times and by deficient modulation of theta and alpha frequencies. Our findings highlight virtual reality (VR) and Neuro-VR as important methodologies for future psychological and gerontological research.

Keywords: ageing; simulated driving; attention; switching costs; neural oscillations; Neuro-VR

1. Introduction

The population of older drivers is rapidly increasing. Identifying age-related changes in cognition that impair driving performance is important to ensure that older adults can continue to drive safely. Although older adults have an overall reduced crash risk compared to young drivers, statistics show that they present a disproportionate risk of at-fault collisions at intersections and collisions caused by a failure to give-way, or to notice other objects, stop signs or signals [1–4]. Older drivers typically report an increased subjective difficulty in processing signs in time [5]. Overall, the evidence suggests that older drivers' collisions are caused by failures in selective attention and switching [6], and that driving skills later in life may be impaired by declines in switching or refocusing attention.

In our recent work [7], we found an age-related decline in the ability to refocus attention from attending to temporally changing events to spatially distributed stimuli. These findings are in line with more general declines in attention with increased age [8–15]. Furthermore, Choi et al. [16] recently reported that reduced attentional control correlates with crash risks during simulated driving, specifically in situations that require a fast resolution of conflicts among competing tasks. Difficulties in switching between temporal events and spatially distributed stimuli [7] could relate to difficulties in switching from attending to the dynamic changes in traffic on the road ahead to attending to road signs and other surrounding objects and events.

In addition to attentional refocusing, the work by Callaghan, Holland and Kessler [7] included an element of goal switching. Participants switched from the goal of identifying a number in a stream of letters to the goal of identifying a letter in a visual search display. Older adults have been shown to find task switching particularly difficult when required to maintain more than one task goal [17]. In the context of driving, one is required to maintain several task goals at once, for example vehicle control, route finding and monitoring traffic, pedestrians and bicycles. Goal switching between different tasks could, therefore, be substantially more challenging for older adults while driving in a dynamically evolving scenario (e.g., [18]).

In the current study, age groups were compared on their ability to switch from allocating attention to dynamic events in time, where participants must attend to the fast changing traffic in front of them (spatially focal), to distributing attention spatially, in order to complete a visual search of a road sign (the target city name "Birmingham" was embedded within 11 other city names, see Figure 1). Participants drove on a simulated dual-carriageway. To maintain immersion, driving events ran on continuously and dynamically from one another, clustering into event pairs or single events, depending on "trial" condition. In Sequential-Task Switch trials, the "road sign visual search" task (2nd task) was preceded by a "braking event" task (1st task), where participants were required to brake in response to a car suddenly driving in front of them from the over-taking lane and braking. Shortly after the braking event, participants were required to refocus their attention spatially in order to complete the "road sign visual search" task. In "Single-Task" trials the "road sign visual search" task (only task) was carried out without a preceding "braking event" task. It was hypothesised that response times (RTs) in response to the road sign would be slower when the sign was preceded by a braking event in Sequential-Task trials, compared to when the sign appeared without a preceding braking event task in Single-Task trials. We refer to this slowing of RTs in Sequential-Task compared to Single-Task trials as Sequential-Task Costs. Crucially, based on our prior work and the existing literature [7,19,20], we hypothesised that there would be an age-related decline in refocusing attention from the temporal to the spatial task while driving, reflected in a greater proportional increase in RTs in Sequential-Task compared to Single-Task trials (Sequential-Task Costs).

Electroencephalography (EEG) was recorded in a subset of 17 participants aged 18–30 and 17 aged 60+ years as a proof-of-principle pilot investigation. Although previous literature has successfully recorded EEG in driving simulator environments [21–23], many studies investigated the effects of fatigue on EEG signals [22,24–26] and only a very limited number of studies have investigated task-related modulations of oscillatory signatures [27–29]. To our knowledge, no study to date has investigated changes to task-related oscillatory signatures in older drivers during naturalistic, simulated driving. We believe this to be an important gap to fill, because oscillatory signatures at alpha, beta and theta frequency have been linked with a variety of attentional and executive functions and their age-related decline, which could be crucial for complex daily activities such as driving.

Based on an extensive body of research on target processing, where increased alpha power is typically thought to reflect inhibition, whereas an alpha desynchronization is thought to reflect enhanced attention [30–36], we expected task-related (Single- vs. Sequential-Task) alpha modulations in relation to the onset of the braking event and subsequently to the appearance of the road sign. Additionally, it was hypothesised that EEG theta power would be modulated by Single- vs. Sequential-Task settings, in accord with previous research linking such modulations with variations in the level of top-down guided attentional control and target processing [37–41].

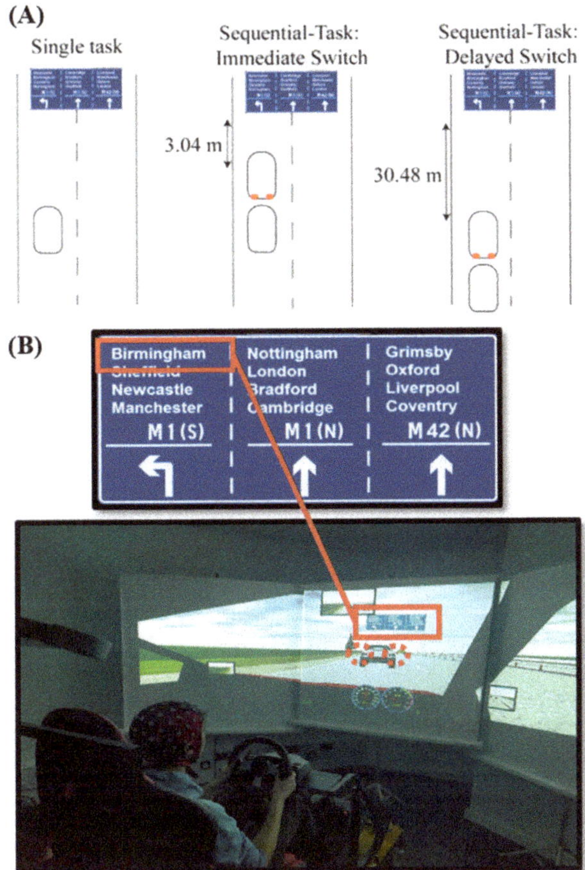

Figure 1. (**A**) Illustration of the three task conditions, including (from left to right) Single-Task, Sequential-Task Immediate Switch and Sequential-Task Delayed Switch. Curved rectangles represent the simulated vehicles of the participant (plain) and in the braking event (with red brake lights illustrated); (**B**) Example of the experimental setup, where the participant is seated in the driving simulator wearing an EEG cap. The projector screen displays the vehicle involved in the "braking event" (1st task; dashed red circles emphasize the brake lights), along with the road sign (2nd task, which always appeared after the braking event) during a Sequential-Task trial. As displayed enlarged at the top, the target word "Birmingham" is listed in the left column of the sign, requiring a speeded left indicator response (the location of the target on the sign varied across trials).

Age-related changes in alpha and theta oscillations have been linked to poorer attention and executive control [42–47]. We, therefore, hypothesised that the older group would show weaker task-related theta and alpha power modulation compared to younger adults. Previous research suggests that age differences in oscillatory patterns during selective attention originate from a fronto-parietal network that has been associated with attentional control [11,48–52]. Indeed, in a magnetoencephalography (MEG) study, Wiesman and Wilson [42] found age effects in oscillatory signatures in a fronto-parietal network during a visuospatial attention task.

However, older drivers may be able to increase the amount of cognitive resources applied during a critical task such as driving, especially if they still regularly engage in this activity, which would be consistent with reports of compensatory recruitment of top-down processes in older age [51,53,54] and would then be reflected in increased and wider-spread task-related modulation of alpha and theta oscillations in the older compared to the younger group.

2. Methods

2.1. Participants

Data from one-hundred and eighteen participants in five age groups (18–30, 40–49, 50–59, 60–69, and 70–91 years) were collected and analysed. A G *Power calculation revealed that a sample of 110 participants was required for detecting a medium effect size (f = 0.35) with 0.9 statistical power. An additional ten participants did not complete the experiment due to simulator sickness, including four participants from the 50–59 years group and two participants from each of the 40–49, 60–69, and 70–91 years groups. All participants had a full driving license, had experience driving in the United Kingdom (UK) and had driven within the last year. Consistent with Callaghan, Holland and Kessler [7], the 18–30 years group were used as a comparison group for age-related cognitive changes for all other groups. The 40–49 and 50–59 years groups provided middle-aged comparison groups for the 60–69 and 70–91 years groups. Age ranges were selected to assess performance not only in older adults, but also in middle age. Participants with photosensitive epilepsy or uncorrected visual impairments were excluded from participation, in addition to those who scored equal to or less than the cut-off for possible cognitive impairment of 87 on the cognitive assessment used, the Addenbrookes Cognitive Examination 3 (ACE-3 [55]).

Participants in the 18–30, 40–49 and 50–59 years groups were recruited from Aston University staff and students and the community. Participants aged over 60 years were recruited from the Aston Research Centre for Healthy Ageing (ARCHA) participation panel. Participants received course credits (students) or a standard fee towards their travel expenses. All participants provided written informed consent before participating. The research was approved by Aston University Research Ethics Committee (approval code #897) and complied with the Declaration of Helsinki.

Two participants from the 70–91 years group scored equal to or lower than the cut-off of 87 on the ACE-3 [55] and were therefore excluded from further analyses. The remaining participants' demographics are presented in Table 1.

Table 1. Participant demographics.

		Age Group (Years)					EEG Group	
		18–30 (*n* = 34)	**40–49 (*n* = 20)**	**50–59 (*n* = 20)**	**60–69 (*n* = 22)**	**70–91 (*n* = 20)**	**18–30 (*n* = 17)**	**60+ (*n* = 17)**
Age (years)	Mean	21.21	43.65	54.75	64.77	75.35	22.88	70.12
	SD	3.36	2.93	2.40	2.86	4.40	4.04	5.20
Sex	Male	10	8	6	14	9	4	10
	Female	24	12	14	8	11	13	7
Handedness	Right	30	17	18	20	20	15	16
	Left	4	2	2	2	0	2	1
ACE-3	Mean	N/A	N/A	96.60	96.18	95.21	N/A	95.29
	SD	N/A	N/A	2.50	2.44	2.49	N/A	2.64

The mean age of each age group, the number of male/female participants, and the number of left/right handed participants for each age group. Mean ACE-3 scores are presented for the 50–59, 60–69, and 70–91 years groups. Handedness data is missing for one participant in the 40–49 years group. Electroencephalography (EEG); Addenbrookes Cognitive Examination 3 (ACE-3); not applicable (N/A); number of participants (*n*); standard deviation (SD).

2.2. Materials and Procedures

2.2.1. Driving Simulator

Participants completed a driving simulator task where they switched from allocating their attention to events changing in time (i.e., temporal attention), as they attended to the fast changing traffic on the road ahead, to distributing attention spatially, in order to complete a visual search of a road sign.

An example of the experimental setup is displayed in Figure 1. Participants were seated comfortably in an adjustable GT Omega Art Racing Simulator Cockpit (RS6 seat), complete with a Logitech G27 Force feedback wheel and pedal set, which incorporated a steering wheel, gear stick, clutch, brake and accelerator pedals. The indicators were paddles to the left and right of the steering wheel that participants could pull towards them to turn on and off. A manual gearstick was to the left of the participant and was programmed to go up to 5th gear. Driving simulator software STISIM Drive™ by Systems Technology Inc. (kernel build: Build 2.10.09) was used to record driving simulator data and to render the driving simulations, which were projected at a resolution of 1280 × 1024 pixels onto three 1.30 × 2.27 m projection screens at a refresh rate of 75 Hz. Data were sampled at a frequency of 60 Hz. The central projection screen was positioned facing the driving seat 2.20 m away from the participant. To fabricate the perception of movement through 3D space, the two peripheral projection screens were positioned adjacent to the central screen, rotated 40 degrees away from the central screen towards the driving seat. The projection included a dashboard which contained a speedometer displaying miles per hour (mph) and a rev counter displaying revolutions per minute. The driving seat was surrounded by a speaker system through which engine and braking sound-effects were produced.

Participants drove on a simulated dual-carriageway. To maintain immersion in the driving simulation study "trials" ran on continuously from one another, where certain events, which occurred either in pairs or in isolation within the ongoing driving scenario, were considered as trials. In "Sequential-Task" trials, participants were required to brake in response to a car suddenly pulling in front of them from the over-taking lane. Participants were instructed to brake as quickly as possible and RTs to brake initiation were recorded (see Figure S1). Shortly after this "braking event", participants were required to refocus their attention spatially to complete a visual search of a road sign, which appeared in front of the driver after they had travelled either another 3.04 m (Immediate Switch condition: ~0.14 s, assuming they are driving at the speed limit) or another 30.48 m (Delayed Switch condition: ~1.42 s, assuming they are driving at the speed limit) after the braking event. The Delayed Switch condition was included to prevent participants predicting exactly when the road sign would appear. In "Single-Task" trials the sign appeared without a preceding braking event. In Sequential-Task trials, the delay between the braking event and the road sign was specified in distance, resulting in the time delay between the braking event and the road sign varying with participants' driving speeds. Note that the Single-Task condition still involved an attention switching element, where participants attended to the road ahead before switching to attend to the road sign. However, Sequential-Task trials were expected to require a heightened effort, both to refocus attention due to a greater enhancement of attention towards temporal events (i.e., towards braking), and to switch from the task goals of braking in response to the braking event to the task goals of indicating in response to the road sign.

An example of the road sign is displayed in Figure 1 (Panel B, top). When the road sign appeared, participants were required to identify the location of the target word "Birmingham" and indicate left if it was in the left column, to signal that they would exit the dual-carriageway, and right if it was in either the middle column or the right hand column to signal that they would stay on the dual-carriageway. Participants were instructed to indicate as quickly and accurately as possible and indicator RTs and accuracy were recorded. The speed at which participants were travelling when the sign appeared was also recorded (see Figure S2). Participants were instructed that the speed limit was 70 mph (consistent with the speed limit on UK dual-carriageways). The maximum speed that participants could drive was programmed at 80 mph.

The city name "Birmingham" was chosen as a visual search target word, as this was the city where the study took place and was therefore familiar to all participants. The target was embedded among 11 distractor names. The stimuli remained the same on all trials and only the order of the stimuli on the sign differed across trials. Distractor stimuli were UK city names. To avoid advantaging participants who were more familiar with certain roads than others, the order of the names on the sign was random. Participants were informed that signs did not represent realistic road signs with regards to the order of names. On 50% of trials the target was in the left column and on 50% of the trials the target was in one of the two right-hand columns, so that 50% of trials required a left-hand indicator response and 50% of trials required a right-hand indicator response (see above for details). There were 36 trials in total and trials were divided into three blocks of 12 trials to provide opportunities for breaks. There were 12 trials in each condition that were pseudorandomised throughout the three blocks. The order of trials in each block can be found in Table S1. Each block took approximately 15 min, depending on the speed at which participants were driving. Due to the length of the experiment (45 min), it was not plausible to increase the number of trials due to concerns about the effects of fatigue on the performance [56]. Fatigue has been shown to effect neural responses recorded with EEG in the driving simulator [24]. Furthermore, a small number of trials per condition maintains the naturalistic nature of the experiment, by preventing over-repetition and minimising the chance to develop artificial strategies that would not be present when responding to surprising events in everyday on-road driving. There are examples across multiple cognitive domains where increasing the number of trials weakens power due to effects of learning throughout the experiment, e.g., [57]. Minimising the number of trials prevents such learning from taking place and better reflects everyday driving performance.

Prior to beginning the task participants took part in two practice driving scenarios. In the first scenario, participants were given the opportunity to familiarise themselves with the controls of the driving simulator while driving around a virtual town. Participants continued driving in the town scenario until they felt confident with the controls, particularly with changing gear, steering and braking. The aim of the second practice scenario was to familiarise participants with the task instructions. Participants completed six practice trials of the task; however, trials differed from experimental trials as they contained no traffic on the road and so there were no braking events and no Sequential-Task switching element to the task.

2.2.2. EEG Acquisition

From the 118 participants, we recorded EEG in 17 participants aged 18–30 years (mean = 22.88 years, SD = 4.05) and in 17 participants aged 60+ years (mean = 70.12 years, SD = 5.20) while they completed the driving simulator task. EEG was recorded with a 64 channel eego™ sports mobile EEG system (ANT Neuro, Enschede, The Netherlands) and digitised at a sampling rate of 500 Hz. Sensors were Ag/AgCl electrodes arranged in accordance with the International 10–10 system. Electrode CPz was taken as an online reference electrode and the ground electrode was positioned at AFz. Participants were instructed to keep their face as relaxed as possible throughout the recording and to keep their head movements to the minimum necessary while driving to minimise muscle artefacts.

3. Data Analysis

3.1. Driving Simulator Task RTs and Accuracy

In the driving simulator task, participants' median indicator RTs on trials where they had both braked successfully in the braking event and indicated correctly in response to the road sign were extracted from raw driving simulator outputs. The proportion of correct indicator responses, braking responses and participants' braking RTs (see Figure S1) and median driving speeds (mph; Figure S2) when passing the road sign were also recorded.

Differences in median indicator RTs between event conditions and age groups were analysed in a 3×5 mixed ANOVA, where event condition (Immediate Switch/Delayed Switch/Single-Task) was a within-subjects factor and age group (18–30/40–49/50–59/60–69/70–91 years) was a between-subjects factor. Multiple comparisons were corrected for with Bonferroni correction.

The data were expected to violate assumptions of equality of variance due to increases in inter-individual variability with age [58,59], yet, there is evidence to support that the ANOVA is robust to violations to homogeneity of variance [60,61]. Where Mauchly's Test of Sphericity was significant, indicating that the assumption of sphericity had been violated, Greenhouse–Geisser corrected statistics were reported.

To interpret the age group × event condition interaction that was identified in the indicator RT ANOVA, percentage differences from the Single-Task condition to each of the Sequential-Task conditions were calculated as measures of "Sequential-Task Costs" for each individual, and independent t-tests were implemented to compare age groups' Sequential-Task Costs. As interaction effects were already shown to be statistically significant in the ANOVA, Restricted Fisher's Least Significant Difference test was applied and corrections for multiple comparisons were not conducted [62]. Where Levene's test for equality of variance was significant ($p < 0.05$), "Equality of variance not assumed" statistics were reported.

3.2. EEG Analysis

3.2.1. Preprocessing

EEG data were read into the Matlab2017a® toolbox Fieldtrip version 20151004 and analysed with version 20161031 [63], bandpass filtered between 0.5–36.0 Hz and epoched from −7.00–3.00 s, where 0.00 s corresponded to the onset of the road sign. Trials were visually inspected for artefacts and trials with large artefacts were removed in addition to trials where participants failed to brake successfully in the braking event or indicate correctly in response to the road sign (Single-Task mean = 1.35 trials, SD = 1.25; Immediate Switch mean = 0.62 trials, SD = 0.78; Delayed Switch mean = 2.50 trials, SD = 0.86). Prior to analysis, independent component analysis was implemented and components with eye-blink or heartbeat signatures were omitted.

3.2.2. EEG Sensor Level Analysis

Noisy sensors were interpolated with the averaged signal from neighbouring electrodes. Time-frequency analysis was carried out by applying a Hanning taper from 2–30 Hz (for every 1 Hz), with five cycles per time-window in steps of 50 ms. For each participant, trials were averaged within each condition (Single-Task/Immediate Switch/Delayed Switch). No baseline correction was applied due to potential group differences in baseline cognitive states. Instead, conditions were compared directly.

3.2.3. Exploratory EEG Source Analysis

In order to explore possible cortical sources for the oscillatory effects that were evident at the sensor-level, and to explore oscillatory signatures while using spatial filtering to suppress noise external to the cortex, sources of theta (3–7 Hz; 0.00–0.80 s) and alpha (8–12 Hz; 1.0–2.0 s) oscillations were localised with exact Low-Resolution Electromagnetic Tomography (eLORETA; Pascual-Marqui, 2007), which has been shown to be less susceptible to noise than LORETA [64,65]. Time-frequency tiles were selected based on the average of time-frequency representations (TFRs) across all conditions and groups (see Figure S3), in order to analyse possible cortical origins of sensor level effects after the onset of the road sign. Note that the TFR in Figure S3 displays power difference in relation to a baseline that was not used in any statistical analysis, collapsed across conditions and groups, and so does not constitute as "double-dipping". Noisy electrodes were excluded prior to re-referencing data to the average of all remaining electrodes. Data were bandpass filtered and epoched to selected time-frequency tiles (detailed above). To avoid spectral leakage of the theta response (e.g. Figure S3), a time window of 1.00–2.00 s was selected for alpha source power analysis. A generic Boundary Element Method head-model was created from a template T1 weighted MRI. Head-models were normalised to MNI space (Montreal Neurological Institute template). Consistent with [66], source model voxels were 5 mm in size to improve the fit near the surface. Covariance matrices (computed for data pooled across conditions) were combined with estimated leadfields to produce common spatial filters. These spatial filters were subsequently applied to data from each condition separately.

3.2.4. Statistical Analysis

The same statistical analysis procedure was followed for sensor and source level analysis in order to explore consistent effects across participants. Two-tailed dependent t-tests were carried out to compare each of the Sequential-Task conditions (Immediate Switch/Delayed Switch) with the Single-Task condition separately for each age group. Multiple comparisons were corrected for with non-parametric cluster permutations [67,68], with 2000 permutations (cluster alpha = $p < 0.05$). Second level analysis was carried out by comparing Sequential-Task Costs at the group level [69,70]. For each participant, the Single-Task condition was subtracted from each of the Sequential-Task conditions separately. These differences were entered into two-tailed independent cluster permutation t-tests (2000 permutations; cluster alpha = $p < 0.05$) to compare age groups. This statistical approach is consistent with recommendations on the Fieldtrip website "How to test an interaction effect using cluster-based permutation tests?" [71] and has been implemented in previous work [69,70].

4. Results

4.1. Driving Simulator Task

All age groups achieved greater than 98% accuracy to respond to the braking event and greater than 95% accuracy to respond to the road sign. No further analysis was conducted on accuracy data.

Indicator RTs

Group means of participants' median indicator RTs in response to the road sign visual search are presented in Figure 2. (Braking RTs are displayed in Figure S1; alongside driving speeds when passing the road sign in Figure S2).

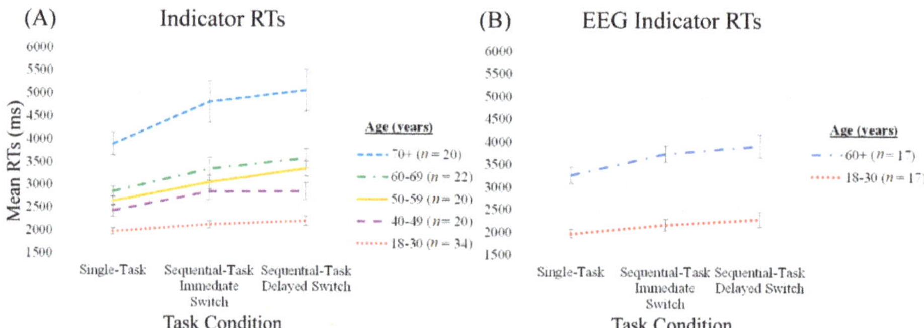

Figure 2. Group means of participants' median indicator RTs in all participants (**A**) and in the subset of participants with EEG recordings (**B**). Note that the data subset shown in B is included in the overall data shown in A and in all analyses conducted with this data. Vertical bars represent the standard error of the mean.

To investigate the hypothesis that Sequential-Task Costs would be greater in older compared to younger groups, a 3 × 5 (event condition × age group) ANOVA was conducted on participants' indicator RTs. There was a significant main effect of age (F (4, 111) = 25.64, $p < 0.001$, $\eta^2_p = 0.48$) and event condition (F (1.66, 185.04) = 57.42, $p < 0.001$, $\eta^2_p = 0.34$) on indicator RTs and a significant age ×

event condition interaction (F (6.67, 185.04) = 3.74, $p = 0.001$, $\eta^2_p = 0.12$). The main effect of age resulted from significantly faster indicator RTs in the 18–30 years group compared to the 50–59 ($p = 0.004$), 60–69 ($p < 0.001$) and 70–91 ($p < 0.001$) years groups, and significantly slower indicator RTs in the 70–91 years group compared to all other groups ($p < 0.001$). There were no other significant age group differences in indicator RTs ($p > 0.10$). The main effect of event condition resulted from significantly faster indicator RTs in the Single-Task condition compared to both Sequential-Task Switch conditions ($p < 0.001$) and faster RTs in the Immediate Switch condition compared to the Delayed Switch condition ($p < 0.001$).

To investigate the hypothesis that age-related increases in Sequential-Task Costs would be seen while driving, the interaction between age and task condition was further explored. Each participant's percentage increase in RTs from the Single-Task condition to the Immediate Switch condition (Immediate Sequential-Task Costs) and from the Single-Task condition to the Delayed Switch condition (Delayed Sequential-Task Costs) were calculated as measures of Sequential-Task Costs. Sequential-Task Costs were entered into independent t-tests to compare groups. T-tests were conducted to interpret the significant age group × event condition interaction and multiple comparisons were not corrected for (see Methods Section 3.1 and [62]). The means and SDs of each group's Immediate and Delayed Sequential-Task Costs are presented in Table 2.

Table 2. Means and SDs of Sequential-Task Costs for each age group.

		Age Group (Years)					**EEG subgroup**	
		18–30 (n = 34)	**40–49 (n = 20)**	**50–59 (n = 20)**	**60–69 (n = 22)**	**70–91 (n = 20)**	**18–30 (n = 17)**	**60+ (n = 17)**
Immediate S-T Costs	Mean	8.62	17.68	15.21	18.71	22.76	10.38	16.98
	SD	16.50	21.34	11.56	29.34	24.35	15.55	15.89
Delayed S-T Costs	Mean	12.67	18.59	26.81	27.58	30.50	15.97	21.45
	SD	20.30	20.31	14.31	24.28	32.37	23.72	21.17

Sequential-Task (S-T) Costs were calculated as the percentage increase in RT from the Single-Task condition to each of the Sequential-Task conditions (Immediate Switch/Delayed Switch) separately.

There were significantly higher Immediate Sequential-Task Costs in the 70–91 years group compared to the 18–30 years group (t (52) = −2.54, $p = 0.014$). Higher Immediate Sequential-Task Costs in the 40–49 years group compared to the 18–30 years group did not reach significance (t (52) = −1.75, $p = 0.087$). There were no other significant age group differences in Immediate Sequential-Task Costs between any age group ($p > 0.10$).

Compared to the 18–30 years group, there were significantly higher Delayed Sequential-Task Costs in the 50–59 (t (52) = −2.74, $p = 0.008$), 60–69 (t (54) = −2.48, $p = 0.016$) and 70–91 (t (27.95) = −2.22, $p = 0.035$) years groups. No other significant age group differences in Delayed Sequential-Task Costs were found between any age group ($p > 0.10$).

4.2. EEG

Group means of EEG participants' median indicator RTs in response to the road sign visual search are presented in Figure 2 panel B. EEG participants' RTs were included in the statistical analysis outlined in Section 4.1.

4.2.1. Sensor-Level TFRs

The TFRs in Figure 3 display a theta response shortly after the road sign appears, which is consistent with top-down guided attentional control and target processing [37–41]. In the Sequential-Task conditions, there was an earlier theta increase that reflects processing of the braking event, which is initiated approximately −0.90 s relative to the onset of the visual search in the Immediate Switch condition and −2.00 s in the Delayed Switch condition. As the vehicle did not brake in front of the participant until either ~0.14 s prior to the onset of the sign in the Immediate Switch condition or ~1.42 s prior to the onset of the sign in the Delayed Switch condition, the earlier onset of theta modulation likely reflects detection of and attention to the vehicle approaching.

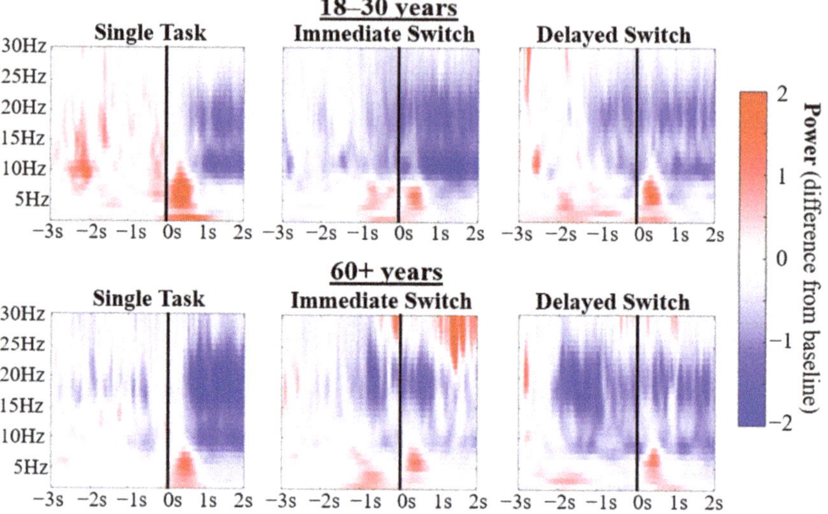

Figure 3. TFRs present power difference from a baseline period of −5.50 s – −3.50 s averaged across a group of 12 anterior electrodes (AF3, AF4, F1, F2, F3, F4, FC1, FC2, FC3, FC4, FC5, FC6). For an average across posterior electrodes see Figure S4 and across all electrodes see Figure S5. Black lines placed over TFRs signify the onset of the road sign at 0.00 s. In the Sequential-Task conditions the car pulled in front of the participant at either 3.04 m (~0.14 s; Immediate Switch) or 30.48 m (~1.42 s; Delayed Switch) prior to the onset of the road sign.

Figure 3 also illustrates an alpha power decrease in a late time window (starting around 0.50 s) in response to the onset of the road sign, consistent with enhanced attention to significant stimuli [30–36].

An early beta (15–25 Hz) decrease is also evident, which was initiated in response to the braking event and maintained throughout the visual search of the road sign. This could reflect enhanced attention or motor preparation as participants learned that a road sign would follow the braking event [72–77]. It appears that this is greater in the older compared to younger group. However, no further analysis was conducted on this, due to concerns about interference from muscle artefacts.

The naturalistic setting of the experiment meant that there was no suitable baseline period to statistically compare relative changes in power in response to task events that appear to be present in Figure 3. Instead, conditions were contrasted directly to compare power across conditions and age groups. To test the hypotheses that there would be age-related decreases in task-related theta and alpha modulation, frequency bands for theta (3–7 Hz) and alpha (8–12 Hz) were selected for further analysis.

4.2.2. Single-Task vs. Sequential-Task Conditions

Figure 4 presents theta (3–7 Hz) and alpha (8–12 Hz) EEG power sensor level effects when contrasting Immediate Switch and Single-Task conditions in each age group (Panels A and C) and when contrasting Delayed Switch and Single-Task conditions in each age group (Panels B and D). No significant effects were found when investigating the event condition × age interactions ($p > 0.10$).

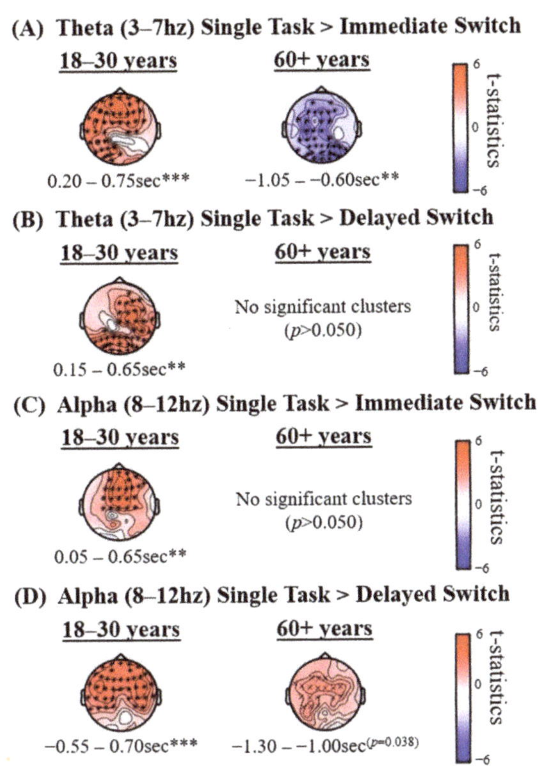

Figure 4. Effects in theta (3–7 Hz) and alpha (8–12 Hz) EEG power when contrasting Immediate Switch and Single-Task conditions in each age group (**A**,**C**), and contrasting Delayed Switch and Single-Task conditions in each age group (**B**,**D**). Topographical plots present *t*-statistics. Cluster significance levels from a two-tailed test are indicated as ** $p < 0.01$, *** $p < 0.001$.

The 18–30 years group but not the 60+ years group displayed significantly higher sensor level alpha power in the Single-Task compared to Sequential-Task conditions (in both Delayed and Immediate Switch conditions). Group comparisons of differences were not significant ($p > 0.10$).

The younger but not the older group displayed significantly higher sensor level theta power in the Single-Task condition compared to both the Immediate Sequential-Task and Delayed Sequential-Task conditions, the cluster for which peaked after the road sign onset. The 60+ years group additionally showed weaker theta power in the Single-Task compared to Sequential-Task Immediate Switch condition in a time window preceding the road sign onset. This likely reflects increased processing in response to the temporal braking event before the onset of the sign in the Immediate Switch condition. Group comparisons of differences were not significant ($p > 0.10$).

4.2.3. Exploratory Source Analysis

As described in Methods, we employed eLORETA to explore the plausible cortical origins of the condition effects found at sensor level (shown in Figure 4) in the period right after the onset of the road sign, and to explore the possible oscillatory signatures while suppressing noise through spatial filtering. The results of these exploratory source analyses could be used as a basis for future confirmatory research.

Consistent with the sensor level analysis, Figure 5 displays higher frontal theta and alpha source power in the Single-Task compared to Sequential-Task conditions in the younger group (Panels A–D, left column), lower posterior alpha power in the Single-Task compared to Immediate (non-significant) and Delayed Switch conditions in the older group (Panel C–D, middle column), and higher frontal theta power in the Single-Task compared to Immediate (non-significant) and Delayed Switch condition in the older group (Panel A–B, middle column). Source analysis highlighted additional effects in the older group after road sign onset (Panels B–C, middle column) and at second level (group comparison) analysis (right column) that were not evident in the sensor level analysis, providing preliminary evidence that spatial filtering may be a useful tool to increase sensitivity in noisy, naturalistic environments. The results presented in Figure 5 suggest that the lower posterior theta power and higher frontal alpha power effects that occur before road sign onset in the older group, evident in Figure 4A,D, may actually have been present in the Single-Task compared to both Sequential-Task conditions, rather than in only the Immediate or Delayed Switch conditions, as Figure 4B,C (right column) would suggest. However, source analysis focused on the period after the onset of the sign, and thus, did not cover the period before onset, where the older group had displayed stronger theta in Sequential-Task conditions compared to Single-Task (Figure 3, bottom and Figure 4A, right).

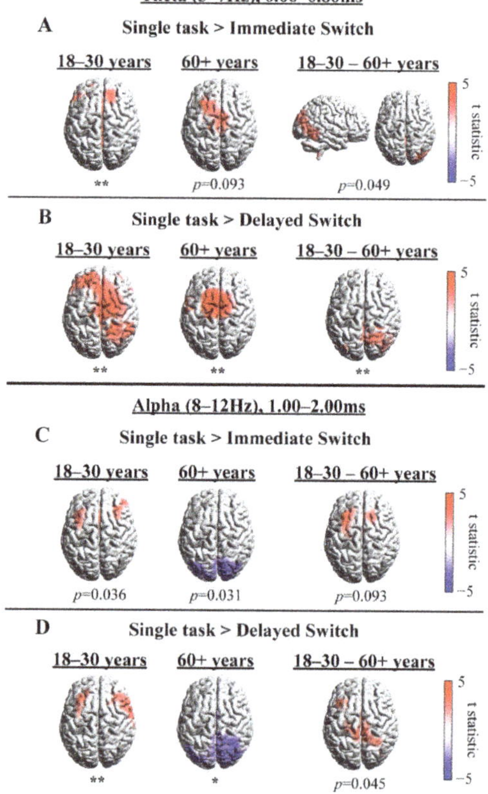

Figure 5. eLORETA source solutions for each age group (left and middle column) and for between-group comparisons (right column). Panels **A** and **B** depict solutions for theta (3–7 Hz; 0.00–0.80 s) and Panels **C** and **D** for alpha (8–12 Hz; 1.0–2.0 s), each for a time period after the onset of the road sign. Panels **A** and **C** show results for the Immediate Switch condition and Panels **B** and **D** for the Delayed Switch condition; both in relation to the Single-Task condition. Cluster significance levels from a two-tailed test are indicated as * $p < 0.025$, ** $p < 0.01$.

5. Discussion

The aim of the study was to investigate whether age-related difficulties in switching from a temporal to a spatial attention task [7] can also be observed in a more naturalistic and ecologically valid setting, i.e., during simulated driving, where older and younger drivers were required to switch from a spatially focal yet temporally complex task (braking due to traffic ahead) to a spatially more distributed task (reading a motorway road sign). EEG was recorded in a subset of participants (17 younger and 17 older drivers) while they completed the driving simulator task, permitting the investigation of oscillatory signatures at theta and alpha frequencies. Our primary hypothesis, that there would be greater Switch Costs in older compared to younger age groups, was supported.

5.1. Response Times

Consistent with hypotheses, RTs to indicate were slower with increased age, in line with general age-related slowing of RTs [8–12]. As hypothesised, RTs to complete the visual search of the road sign were significantly slower in the two Sequential-Task conditions compared to the Single-Task condition, reflecting costs of switching from responding to a traffic event ahead (braking) to responding to a spatially distributed motorway road sign (signalling). In other words, it was more difficult to broaden the focus of attention to distribute attention spatially (to an overhead motorway road sign, in order to find a target city name among distractors) when attention mechanisms were already engaged in responding to a temporally complex event (the car braking in front) in the Sequential-Task conditions.

RTs in the Delayed Switch condition were significantly slower than in the Immediate Switch condition. This was unexpected, as faster RTs were predicted when participants had more time to refocus their attention spatially after the braking event. In our previous laboratory-based experiment [7], participants were faster on the visual search task when they could prepare to switch sooner. It is likely that participants learned that the road sign followed the braking event. After initially being a distraction, the braking event may have served as a cue for the motorway sign. There is a significant body of literature that demonstrates that temporal orienting of attention is enhanced at shorter (compared to longer) time intervals between a cue and a target, and RTs to detect a target decrease if the time of stimulus onset is predictable [50,78–80]. When the distance between the braking event and the road sign onset is farther in the Delayed Switch condition, the variability in the time of the road sign onset increases, due to variability in the speed that participants are driving (see also Figure S2). Increased RTs in the Delayed Switch condition compared to the Immediate Switch condition were therefore likely due to variability in the onset time of the road sign (making it less predictable), combined with a longer time period between the braking event and the road sign onset. There was no such variability in the onset time of the visual search task in Callaghan, Holland and Kessler [7], and so RTs were faster when participants had more time to prepare to switch. In addition, it could be that at long delays, participants focused their attention back on the road traffic and therefore had to re-focus attention spatially when the sign appeared, which shares some resemblance with the classic Inhibition-of-Return effect [81,82]. However, possible explanations of the cognitive mechanisms underlying longer RTs in the Delayed compared to Immediate Switch condition are merely speculative and would benefit from further research that could include eye-tracking, for instance. Such considerations also highlight the complex issues that have to be considered in dynamic naturalistic settings (e.g., [18]).

The hypothesis that there would be an age-related decline in switching attention while driving was supported. There was a greater increase in RTs from the Single-Task condition to the Immediate Switch condition in the 40–49 and 70–91 years groups compared to the 18–30 years group (although this did not reach significance in the 40–49 years group). Greater Delayed Sequential-Task Costs were found in the 50–59, 60–69 and 70–91 years groups compared to the 18–30 years group. These findings demonstrate that age-related declines in refocusing attention between tasks are not only observed in a standard computer-based paradigm [7] but are also present in more naturalistic settings, such as simulated driving.

Findings of significantly increased Delayed Sequential-Task Costs but not Immediate Sequential-Task Costs in the 50–59 and 60–69 years groups signifies that these age groups find switching easier when the onset time of the stimulus is after a short delay (~0.14 s) and predictable, but are more impaired when the time of stimulus onset is after a longer delay (~1.42 ms) and ambiguous—which is arguably more typical for real-life driving. These findings are in line with older age groups relying more on top-down guidance to control attention, such as implementing the use of temporal cues [83–85], as well as evidence towards impaired anticipatory attention mechanisms [86–88]. Furthermore, there is evidence to suggest that preparatory processes during task switching function well in older age and that some performance differences may result from age-related changes in the strategies employed for task implementation [89]. It is also important to highlight that older drivers were driving more slowly after braking, at the point of road sign onset, than the youngest group (Figure S2). This resulted in a slightly longer delay of road sign onset in older drivers, which may have precluded increased Immediate Sequential-Task Costs in the 50–59 and 60–69 years groups. In contrast, with a longer and even more speed-dependent delay of road sign onset in the Delayed Switch condition, attention may have been fully re-directed to the road traffic, thereby preventing any benefit from temporal cueing strategies used to anticipate the road sign. Finally, a contributing factor for the lack of significantly higher Immediate Sequential-Task Costs in the 50–59 and 60–69 years groups could be age-related increased variability in RTs, generally, and in Sequential-Task Costs more specifically.

In contrast to the 50–59 and 60–69 years groups, participants aged 40–49 years displayed higher Immediate Sequential-Task Costs but no difference in Delayed Sequential-Task Costs compared to participants aged 18–30 years. Note that Figure 2 displays little difference between Immediate and Delayed Switch RTs in the 40–49 years group, and so the discrepancies in age group differences between Immediate and Delayed Sequential-Task Costs are partly driven by higher Delayed compared to Immediate Sequential-Task Costs in the 18–30 years group (evident in Table 2).

5.2. EEG

The pattern of RT results was replicated in principle for the subset of 34 participants (17 younger and 17 older participants) for which EEG was recorded during simulated driving (Figure 2). EEG offered an online indicator of brain responses in relation to dynamically unravelling traffic events on-screen.

On inspection of Figure 3 (also Figures S3–S5), theta power appeared to increase shortly after the road sign onset, consistent with the notion of theta involvement in top-down guided attentional control and visual search processing [37–41]. A later alpha desynchronization was apparent after road sign onset, likely related to increased attention to the road sign [30–36]. An early beta decrease was also seen, which was initiated in response to the braking event, and was maintained throughout the visual search of the road sign. It is likely that this reflects enhanced attention or motor preparation as participants learned that a road sign (requiring a response) would follow the braking event [72–77]. However, no further analysis was conducted on beta, due to concerns of interference from muscle artefacts due to active driving movements. Figure S5 demonstrates that this was not a problem for theta and alpha frequencies. Furthermore, relative changes in power in response to task events, which appear to be present in Figure 3, were not tested statistically, as the naturalistic, dynamic and continuous nature of the experimental setup meant there was no suitable baseline period. Instead, conditions were compared directly.

The hypothesis that there would be a weaker theta response in the older compared to younger group was supported. In a time window after the onset of the road sign, the 18–30 years group displayed higher theta in the Single-Task condition compared to each of the Sequential-Task conditions. In contrast, the 60+ years group showed no significant difference between Single-Task and Sequential-Task conditions in sensor level theta power in response to the road sign. Interestingly, stronger theta power was observed for the older group before the onset of the road sign in the Immediate Sequential-Task condition (compared to the Single-Task condition), which was not observed for the younger group, and which could indicate a greater use of top-down, cue-based (braking event) strategies that has previously been reported in older adults [83–85]. However, when directly compared, group differences did not reach significance at sensor level. In contrast, group comparisons in source space indicated reduced posterior theta in older drivers (Figure 5) in response to the road sign, which reached statistical significance in the Delayed Switch contrast. This posterior reduction is in accord with recent MEG findings [90] using Callaghan, Holland and Kessler's [7] rapid serial visual presentation (RSVP) – visual search paradigm.

Greater theta power in the Single-Task condition compared to Sequential-Task conditions is an interesting finding. It could be that there are fewer attentional resources available to process the sign in the Sequential-Task conditions after attending to the braking event. Alternatively, higher theta power in the Single-Task condition could reflect enhanced cognitive effort to attend to the road sign when driving at high speeds, in contrast to when participants brake shortly before the road sign onset in response to the braking event in the Sequential-Task conditions. If the latter hypothesis is correct, the theta power observed in the younger group could reflect enhanced attentional resources directed to the road sign (more so in the Single-Task compared to Sequential-Task condition), which contrasts from the older group, who drive at slower speeds to read the road sign. In addition, only the older drivers increased their theta after the braking event in anticipation of the road sign (Figure 3, bottom; Figure 4A, right), potentially using the braking event as a cue. Such a difference in strategy is further consistent with evidence showing that older drivers adjust their driving behaviour to compensate for slowed RTs [91]. Such adjustments to behaviour were also observed in the current study in participants' driving speeds (Figure S2). Additionally, the posterior distribution of group differences in theta power in response to the road sign (see Figure 5; theta modulation was greater in younger than older drivers) could further corroborate a visual attention deficit in older drivers, resulting in slowed RTs and the urge to slow down. Such an interpretation is further corroborated by recent MEG findings [90] using the Callaghan, Holland and Kessler [7] paradigm.

The hypothesis that there would be age group differences in alpha modulation was qualitatively supported, although group differences did not reach significance. As shown in Figures 3–5, the 18–30 years but not the 60+ years group displayed greater alpha power in the Single-Task condition compared to the Sequential-Task conditions (in both Delayed and Immediate Switch conditions). Figure 5 shows that older drivers instead display lower posterior alpha power in the Single- compared to Sequential-Task conditions. Alpha oscillations are typically thought to relate to attention, with decreased power reflecting enhanced attention [30–36]. The younger group therefore seem to demonstrate enhanced attention towards the road sign in the Sequential-Task compared to Single-Task conditions, which could allow for faster re-distribution of spatial attention, compatible with the RT results. Older adults, in contrast, show lower posterior alpha source power in the Single-Task condition compared to Sequential-Task conditions, which could reflect enhanced visual processing during undisturbed driving (no braking required). This interpretation is compatible with alpha effects observed during the Callaghan, Holland and Kessler [7] task, using MEG [90]. Enhanced attention to the road sign in the Sequential-Task (compared to Single-Task) conditions in the younger but not older group further supports the notion that younger drivers adjust well and quickly to dynamic changes in the driving environment, enabling them to drive at higher speeds, while older adults drive more slowly in order to cope with their reduced attentional flexibility.

5.3. Limitations and Future Work

There were a number of limitations to the study. Firstly, we only investigated conditions where a temporal event preceded a spatial event, and not vice versa, in order to remain consistent with Callaghan, Holland and Kessler [7] and to contain the number of conditions. This means that only inferences about switching from temporal to spatial attention can be drawn. Due to the increased length of naturalistic trials, including additional conditions was not feasible. Further research is needed to explore whether similar age-related changes are present when switching from a spatial to a temporal attention task. Based on previous literature showing that older adults find it more difficult to narrow their focus of attention from two RSVP streams to one [92], it could be expected that older adults also find it more difficult to switch from distributing their attention spatially to focusing their attention on temporally changing events (in a single location). However, it could also be that the salience of such temporal events efficiently attracts attention exogenously, outweighing increased cognitive demands.

Secondly, due to the STISIM drive software providing only the option to programme events into the scenario in distance travelled rather than time elapsed, the sign appeared either 3.04 m or 30.48 m after the braking event. This resulted in the temporal dynamics of each trial being affected by the speed that the participant was driving. This has been taken into careful consideration when interpreting findings. It further emphasises the difficulties that researchers encounter when leaving the "safety" of traditional computer experiments and embark on using more naturalistic, dynamic scenarios.

Similarly, to maintain ecological validity, participants were given control over the driving simulator vehicle and were therefore free to brake and accelerate at any time. Although the braking event did not occur until either ~0.14 s prior to the onset of the sign in the Immediate Switch condition or ~1.42 s prior to the onset of the sign in the Delayed Switch condition it is likely the participant detected the vehicle as a possible hazard prior to these time points. Older drivers have been shown to adjust their driving behaviour to compensate for slowed RTs [91] and may therefore have begun to brake at the earliest sign of a possible hazard, affecting measures of braking RT. This would explain the unusual pattern of braking RTs observed across age groups (Figure S1), where the youngest age group display slower braking RTs than older groups. Although the individual group comparisons did not reach significance with a conservative Bonferroni multiple comparisons correction, there was a significant main effect of age ($p = 0.049$). A possible explanation for the interesting pattern in braking RTs could be that younger drivers also tend to drive at higher velocities (see Figure S2), which could inflate their RTs as stimuli are changing more rapidly in the environment. An alternative explanation is that younger participants have less overall driving experience compared to older drivers. More experienced drivers conduct more anticipatory braking responses, whatever their age, e.g., [93]. See also Bao and Boyle [94] for a similar pattern of driving behaviour results, where the middle-aged group spent more time scanning their rear-view mirror compared to both younger and older adults.

A further limitation of the study is that the visual search road sign (displayed in Figure 1) contained three columns of names that the target could occur in, with the left-hand column corresponding to a left indicator response and the two right-hand columns corresponding to a right indicator response. On 50% of trials a right-hand indicator response was required and on 50% of trials a left-hand indicator response was required. This enabled participants to apply a strategy of only reading words in the left column. It is not expected that this affected our main findings for two reasons. Firstly, the number of occurrences of the target word in each column of the sign was matched across conditions, trials were pseudo-randomised within each of the three blocks and the order that the three blocks were completed in was counterbalanced across participants. Therefore, this strategy could not have been systematically implemented in one condition more than the others. Secondly, there is no evidence to suggest that older participants would be less able to identify and implement this strategy compared to younger adults. Conversely, evidence consistently shows that older participants utilise top-down cues more than younger adults, including contextual cues in a realistic visual search [83]. Older age groups displayed slower RTs and higher Dual-Task Costs compared to younger adults, suggesting that, if they were more likely to implement such a strategy, it did not affect our findings.

A common challenge in ageing research is self-selection bias during recruitment, in which volunteers tend to be highly educated, physically and socially active individuals; traits which have all been identified to improve cognitive reserve and protect against cognitive decline [95–98]. Similarly, 75% of the ARCHA participation panel members attending our previous experiments had degree level education or higher, with only three participants from the two older groups (60–69 and 70+ years groups) in Callaghan, Holland and Kessler [7] having lower than A-level equivalent education. A limitation of the current work is that we did not include level of education, physical activity or social activity as covariates in our current analysis. However, as the majority of participants in all age groups are expected to have degree level education or higher, we do not believe level of education can account for group differences in switching performance. On the contrary, it is possible that the current findings underestimate difficulties in switching in the general ageing population.

A further limitation to the current work is that groups were not matched on gender due to the practical challenges of recruiting a large sample of middle and older aged adults. We have no reason to expect that there would be gender differences in the current study. For example, Bao and Boyle [94] found no gender differences in a similar driving simulator task in which the time that participants spent visually scanning their environment was recorded. However, further research is required that directly investigates gender differences to confirm this speculation.

Lastly, the interaction between event condition and age did not reach significance when comparing the EEG subsets of participants (see Supplementary Materials). This is likely due to the small number of participants (17 per group) and high variability (see Table 2). However, the overall pattern of indicator RTs remained the same as the data from all 116 participants (see Figure 2). Future research should to aim to replicate the current EEG findings with larger samples. Such limitations highlight the challenges of measuring brain signals in naturalistic environments. The current work presents an important step in progressing the field of naturalistic neurocognitive ageing research.

6. Conclusions

Virtual reality is an emerging new tool in psychological research that enables fully controlled computerised experiments with significantly increased realism. In contrast to field studies in the real-world, experimental control remains largely with the researchers and any risks imposed by real-world situations (e.g., traffic) are only virtual. Furthermore, we propose that when used in conjunction with EEG and other neuroimaging tools, unprecedented insights could be gained into human processing in realistic hazardous scenarios.

The aim of the current study was to investigate whether age-related difficulties in switching from a temporal to a spatial attention task [7] are also seen in a naturalistic setting, during simulated driving, and whether differences in behavioural performance would map onto neural oscillatory signatures. Consistent with hypotheses, RTs were slower in the two Sequential-Task conditions compared to the Single-Task condition. Unexpectedly, RTs were slower in the Delayed Switch condition compared to the Immediate Switch condition. This is likely to be due to the two well-known phenomena of increased RTs with increased time elapsed between the presentation of a cue (i.e., the braking event) and a target stimulus (i.e., the road sign) and increased RTs when the temporal onset of a target stimulus is unpredictable [50,78–80]. Future (hypothesis driven) work should aim to confirm this speculation.

The hypothesis that there would be an age-related decline in switching from temporal to spatial attention while driving was supported, reflected in greater Sequential-Task Costs in older age groups compared to the youngest age group. These findings support that age-related declines in refocusing from a temporal event to spatially distributed stimuli are not only observed in the computer-based switching task but are also present in naturalistic settings such as when driving. Age-related changes in task-related theta and alpha modulations were found, where older adults showed a weaker theta and alpha modulation compared to younger adults, which may imply that age groups implement different strategies to cope with attentional demands while driving.

Our findings provide a focus for the future development of training interventions to help older drivers to drive safely for longer. Driving cessation is detrimental to an older person's independence and mental health [99–101]. Overall, older drivers compensate for slower RTs and general cognitive decline in their driving behaviour [91]. However, recent findings have shown that protective factors against cognitive decline in older age, such as level of education and engagement, no longer facilitate compensation later in older age, when the cognitive reserve available to compensate is no longer sufficient for the demand for compensation, caused by declines in underlying cognitive resources [102,103]. The development of an assessment and training intervention that aims to assess and improve the ability to switch between tasks during driving, such as attending to traffic and reading road signs, could therefore be beneficial to the ageing population.

Supplementary Materials: The following are available online at http://www.mdpi.com/2076-3425/10/8/530/s1, Table S1: Counterbalanced Conditions; Figure S1: Group means of participants' median braking RTs; Figure S2: Group means of participants' driving speeds when they passed the road sign; Figure S3: TFR in which time-frequency tiles for exploratory source analysis were selected, presenting power difference from a baseline period of −5.50 s–−3.50 s averaged across 12 anterior electrodes and across all conditions and all age groups; Figure S4: TFRs present power in relation to a baseline period of −5.50 s–−3.50 s averaged across a group of posterior electrodes; Figure S5: TFRs present power in relation to a baseline period of −5.50 s–−3.50 s in all electrodes that were included in the analysis; Indicator RT statistics in EEG groups.

Author Contributions: Writing—Review & Editing and Resources, E.H., H.W., C.H. and K.K.; Conceptualization and Methodology, E.H., C.H. and K.K.; Formal Analysis, Visualization, Data Curation, and Project Administration: E.H.; Investigation and Software, E.H., and H.W.; Writing—Original Draft Preparation, E.H. and K.K.; Supervision and Funding Acquisition C.H. and K.K. All authors have read and agreed to the published version of the manuscript.

Funding: This research was supported by funding from The Rees Jeffreys Road Fund, the Royal Automobile Club Foundation, and the School of Life and Health Sciences, Aston University.

Conflicts of Interest: The authors declare no conflict of interest.

References

1. Arai, A.; Arai, Y. Self-assessed driving behaviors associated with age among middle-aged and older adults in Japan. *Arch. Gerontol. Geriatr.* **2015**, *60*, 39–44. [CrossRef] [PubMed]
2. Guo, A.; Brake, J.; Edwards, S.; Blythe, P.; Fairchild, R. The application of in-vehicle systems for elderly drivers. *Eur. Transp. Res. Rev.* **2010**, *2*, 165–174. [CrossRef]
3. Hakamies-Blomqvist, L.E. Fatal accidents of older drivers. *Accid. Anal. Prev.* **1993**, *25*, 19–27. [CrossRef]
4. McGwin, J.G.; Brown, D.B. Characteristics of traffic crashes among young, middle-aged, and older drivers. *Accid. Anal. Prev.* **1999**, *31*, 181–198. [CrossRef]
5. Musselwhite, C.B.A.; Haddad, H. Exploring older drivers' perceptions of driving. *Eur. J. Ageing* **2010**, *7*, 181–188. [CrossRef]
6. Parasuraman, R.; Nestor, P.G. Attention and Driving Skills in Aging and Alzheimer's Disease. *J. Hu. Factors Ergon. Soc.* **1991**, *33*, 539–557. [CrossRef]
7. Callaghan, E.; Holland, C.; Kessler, K. Age-related changes in the ability to switch between temporal and spatial attention. *Front. Aging Neurosci.* **2017**, *9*, 28. [CrossRef]
8. Bennett, I.J.; Motes, M.A.; Rao, N.K.; Rypma, B. White matter tract integrity predicts visual search performance in young and older adults. *Neurobiol. Aging* **2012**, *33*. [CrossRef]
9. Foster, J.K.; Behrmann, M.; Stuss, D.T. Aging and visual search: Generalized cognitive slowing or selective deficit in attention? *Aging Cogn.* **1995**, *2*, 279–299. [CrossRef]
10. Humphrey, D.G.; Kramer, A.F. Age differences in visual search for feature, conjunction, and triple-conjunction targets. *Psychol. Aging* **1997**, *12*, 704–717. [CrossRef]
11. Nagamatsu, L.S.; Munkacsy, M.; Liu-Ambrose, T.; Handy, T.C. Altered visual-spatial attention to task-irrelevant information is associated with falls risk in older adults. *Neuropsychologia* **2013**, *51*, 3025–3032. [CrossRef] [PubMed]
12. Plude, D.J.; Doussard-Roosevelt, J.A. Aging, selective attention and feature integration. *Psychol. Aging* **1989**, *4*, 98–105. [CrossRef] [PubMed]
13. Lee, T.-Y.; Hsieh, S. The Limits of Attention for Visual Perception and Action in Aging. *Aging Neuropsychol. Cogn.* **2009**, *16*, 311–329. [CrossRef] [PubMed]
14. Lahar, C.J.; Isaak, M.I.; McArthur, A.D. Age Differences in the Magnitude of the Attentional Blink. *Aging Neuropsychol. Cogn.* **2001**, *8*, 149–159. [CrossRef]
15. Maciokas, J.B.; Crognale, M.A. Cognitive and attentional changes with age: Evidence from attentional blink deficits. *Exp. Aging Res.* **2003**, *29*, 137–153. [CrossRef]
16. Choi, H.; Kasko, J.; Feng, J. An Attention Assessment for Informing Older Drivers' Crash Risks in Various Hazardous Situations. *Gerontologist* **2019**, *59*, 112–123. [CrossRef]
17. Wasylyshyn, C.; Verhaeghen, P.; Sliwinski, M.J. Aging and task switching: A meta-analysis. *Psychol. Aging* **2011**, *26*, 15–20. [CrossRef]

18. Torrens-Burton, A.; Hanley, C.J.; Wood, R.; Basoudan, N.; Norris, J.E.; Richards, E.; Tales, A. Lacking Pace but Not Precision: Age-Related Information Processing Changes in Response to a Dynamic Attentional Control Task. *Brain Sci.* **2020**, *10*, 390. [CrossRef]
19. Salthouse, T.A. Speed of behavior and its implications for cognition. In *Handbook of the Psychology of Aging*, 2nd ed.; Van Nostrand Reinhold Co.: New York, NY, USA, 1985; pp. 400–426.
20. Birren, J.E. Translations in gerontology: From lab to life: Psychophysiology and speed of response. *Am. Psychol.* **1974**, *29*, 808–815. [CrossRef]
21. Campagne, A.; Pebayle, T.; Muzet, A. Correlation between driving errors and vigilance level: Influence of the driver's age. *Physiol. Behav.* **2004**, *80*, 515–524. [CrossRef]
22. Lowden, A.; Anund, A.; Kecklund, G.; Peters, B.; Akerstedt, T. Wakefulness in young and elderly subjects driving at night in a car simulator. *Accid. Anal. Prev.* **2009**, *41*, 1001–1007. [CrossRef] [PubMed]
23. Ross, V.; Vossen, A.Y.; Smulders, F.T.Y.; Ruiter, R.A.C.; Brijs, T.; Brijs, K.; Wets, G.; Jongen, E.M.M. Measuring working memory load effects on electrophysiological markers of attention orienting during a simulated drive. *Ergonomics* **2018**, *61*, 429–443. [CrossRef] [PubMed]
24. Ahn, S.; Nguyen, T.; Jang, H.; Kim, J.G.; Jun, S.C. Exploring Neuro-Physiological Correlates of Drivers' Mental Fatigue Caused by Sleep Deprivation Using Simultaneous EEG, ECG, and fNIRS Data. *Front. Human Neurosci.* **2016**, *10*. [CrossRef] [PubMed]
25. Hsu, S.H.; Jung, T.P. Monitoring alert and drowsy states by modeling EEG source nonstationarity. *J. Neural Eng.* **2017**, *14*. [CrossRef]
26. Perrier, J.; Jongen, S.; Vuurman, E.; Bocca, M.L.; Ramaekers, J.G.; Vermeeren, A. Driving performance and EEG fluctuations during on-the-road driving following sleep deprivation. *Biol. Psychol.* **2016**, *121*, 1–11. [CrossRef]
27. Huang, R.S.; Jung, T.P.; Makeig, S. Tonic Changes in EEG Power Spectra during Simulated Driving. In *International Conference on Foundations of Augmented Cognition*; Springer: Berlin/Heidelberg, Germany, 2009; pp. 394–403.
28. Vossen, A.Y.; Ross, V.; Jongen, E.M.M.; Ruiter, R.A.C.; Smulders, F.T.Y. Effect of working memory load on electrophysiological markers of visuospatial orienting in a spatial cueing task simulating a traffic situation. *Psychophysiology* **2016**, *53*, 237–251. [CrossRef]
29. Lin, C.T.; Chen, S.A.; Chiu, T.T.; Lin, H.Z.; Ko, L.W. Spatial and temporal EEG dynamics of dual-task driving performance. *J. Neuroeng. Rehabil.* **2011**, *8*. [CrossRef]
30. Klimesch, W.; Sauseng, P.; Hanslmayr, S. EEG alpha oscillations: The inhibition–timing hypothesis. *Brain Res. Rev.* **2007**, *53*, 63–88. [CrossRef]
31. Yamagishi, N.; Callan, D.E.; Goda, N.; Anderson, S.J.; Yoshida, Y.; Kawato, M. Attentional modulation of oscillatory activity in human visual cortex. *Neuroimage* **2003**, *20*, 98–113. [CrossRef]
32. Hanslmayr, S.; Aslan, A.; Staudigl, T.; Klimesch, W.; Herrmann, C.S.; Bäuml, K.-H. Prestimulus oscillations predict visual perception performance between and within subjects. *Neuroimage* **2007**, *37*, 1465–1473. [CrossRef]
33. Hanslmayr, S.; Klimesch, W.; Sauseng, P.; Gruber, W.; Doppelmayr, M.; Freunberger, R.; Pecherstorfer, T. Visual discrimination performance is related to decreased alpha amplitude but increased phase locking. *Neurosci. Lett.* **2005**, *375*, 64–68. [CrossRef] [PubMed]
34. Sauseng, P.; Klimesch, W.; Stadler, W.; Schabus, M.; Doppelmayr, M.; Hanslmayr, S.; Gruber, W.R.; Birbaumer, N. A shift of visual spatial attention is selectively associated with human EEG alpha activity. *Eur. J. Neurosci.* **2005**, *22*, 2917–2926. [CrossRef] [PubMed]
35. Thut, G.; Nietzel, A.; Brandt, S.A.; Pascual-Leone, A. alpha-Band electroencephalographic activity over occipital cortex indexes visuospatial attention bias and predicts visual target detection. *J. Neurosci.* **2006**, *26*, 9494–9502. [CrossRef] [PubMed]
36. Capotosto, P.; Babiloni, C.; Romani, G.L.; Corbetta, M. Frontoparietal Cortex Controls Spatial Attention through Modulation of Anticipatory Alpha Rhythms. *J. Neurosci.* **2009**, *29*, 5863–5872. [CrossRef]
37. Min, B.-K.; Park, H.-J. Task-related modulation of anterior theta and posterior alpha EEG reflects top-down preparation. *BMC Neurosci.* **2010**, *11*, 1–8. [CrossRef]
38. Cavanagh, J.F.; Frank, M.J. Frontal theta as a mechanism for cognitive control. *Trends Cogn. Sci.* **2014**, *18*, 414–421. [CrossRef]

39. Cavanagh, J.F.; Cohen, M.X.; Allen, J.J.B. Prelude to and resolution of an error: EEG phase synchrony reveals cognitive control dynamica during action monitoring. *J. Neurosci.* **2009**, *29*, 98–105. [CrossRef]
40. Demiralp, T.; Başar, E. Theta rhythmicities following expected visual and auditory targets. *Int. J. Psychophysiol.* **1992**, *13*, 147–160. [CrossRef]
41. Green, J.J.; Conder, J.A.; McDonald, J.J. Lateralized frontal activity elicited by attention-directing visual and auditory cues. *Psychophysiology* **2008**, *45*, 579–587. [CrossRef]
42. Wiesman, A.I.; Wilson, T.W. The impact of age and sex on the oscillatory dynamics of visuospatial processing. *Neuroimage* **2019**, *185*, 513–520. [CrossRef]
43. Cummins, T.D.R.; Finnigan, S. Theta power is reduced in healthy cognitive aging. *Int. J. Psychophysiol.* **2007**, *66*, 10–17. [CrossRef] [PubMed]
44. Finnigan, S.; O'Connell, R.G.; Cummins, T.D.R.; Broughton, M.; Robertson, I.H. ERP measures indicate both attention and working memory encoding decrements in aging. *Psychophysiology* **2011**, *48*, 601–611. [CrossRef] [PubMed]
45. Reichert, J.L.; Kober, S.E.; Witte, M.; Neuper, C.; Wood, G. Age-related effects on verbal and visuospatial memory are mediated by theta and alpha II rhythms. *Int. J. Psychophysiol.* **2016**, *99*, 67–78. [CrossRef] [PubMed]
46. Van de Vijver, I.; Cohen, M.X.; Ridderinkhof, K.R. Aging affects medial but not anterior frontal learning-related theta oscillations. *Neurobiol. Aging* **2014**, *35*, 692–704. [CrossRef]
47. Li, L.; Zhao, D.D. Age-Related Inter-Region EEG Coupling Changes During the Control of Bottom-Up and Top-Down Attention. *Front. Aging Neurosci.* **2015**, *7*. [CrossRef]
48. Fu, S.M.; Greenwood, P.M.; Parasuraman, R. Brain mechanisms of involuntary visuospatial attention: An event-related potential study. *Human Brain Mapp.* **2005**, *25*, 378–390. [CrossRef]
49. Gross, J.; Schmitz, F.; Schnitzler, I.; Kessler, K.; Shapiro, K.; Hommel, B.; Schnitzler, A. Modulation of long-range neural synchrony reflects temporal limitations of visual attention in humans. *Proc. Natl. Acad. Sci. USA* **2004**, *101*, 13050–13055. [CrossRef]
50. Coull, J.T.; Nobre, A.C. Where and when to pay attention: The neural systems for directing attention to spatial locations and to time intervals as revealed by both PET and fMRI. *J. Neurosci.* **1998**, *18*, 7426–7435. [CrossRef]
51. Madden, D.J.; Spaniol, J.; Whiting, W.L.; Bucur, B.; Provenzale, J.M.; Cabeza, R.; White, L.E.; Huettel, S.A. Adult age differences in the functional neuroanatomy of visual attention: A combined fMRI and DTI study. *Neurobiol. Aging* **2007**, *28*, 459–476. [CrossRef]
52. Shapiro, K.; Hillstrom, A.P.; Husain, M. Control of visuotemporal attention by inferior parietal and superior temporal cortex. *Curr. Biol.* **2002**, *12*, 1320–1325. [CrossRef]
53. Fabiani, M.; Low, K.A.; Wee, E.; Sable, J.J.; Gratton, G. Reduced Suppression or Labile Memory? Mechanisms of Inefficient Filtering of Irrelevant Information in Older Adults. *J. Cogn. Neurosci.* **2006**, *18*, 637–650. [CrossRef] [PubMed]
54. Davis, S.W.; Dennis, N.A.; Daselaar, S.M.; Fleck, M.S.; Cabeza, R. Qué PASA? The Posterior–Anterior Shift in Aging. *Cereb. Cortex* **2008**, *18*, 1201–1209. [CrossRef] [PubMed]
55. Noone, P. Addenbrooke's Cognitive Examination-III. *Occup. Med.* **2015**, *65*, 418–420. [CrossRef] [PubMed]
56. Liu, Y.C.; Wu, T.J. Fatigued driver's driving behavior and cognitive task performance: Effects of road environments and road environment changes. *Saf. Sci.* **2009**, *47*, 1083–1089. [CrossRef]
57. Jaeger, T.F.; Burchill, Z.; Bushong, W. Strong Evidence for Expectation Adaptation during Language Understanding, not a Replication Failure. A Reply to Harrington Stack, James, and Watson (2018). Available online: https://osf.io/4vxyp/ (accessed on 15 February 2019).
58. Hale, S.; Myerson, J.; Smith, G.A.; Poon, L.W. Age, variability, and speed: Between-subjects diversity. *Psychol. Aging* **1988**, *3*, 407. [CrossRef]
59. Morse, C.K. Does variability increase with age? An archival study of cognitive measures. *Psychol. Aging* **1993**, *8*, 156. [CrossRef]
60. Budescu, D.V. The power of the F test in normal populations with heterogeneous variances. *Educ. Psychol. Meas.* **1982**, *42*, 409–416. [CrossRef]
61. Budescu, D.V.; Appelbaum, M.I. Variance stabilizing transformations and the power of the F test. *J. Educ. Behav. Stat.* **1981**, *6*, 55–74. [CrossRef]
62. Snedecor, G.W.; Cochran, W. *Statistical Methods*; Iowa State University: Ames, IA, USA, 1967; p. 593.

63. Oostenveld, R.; Fries, P.; Maris, E.; Schoffelen, J.-M. FieldTrip: Open source software for advanced analysis of MEG, EEG, and invasive electrophysiological data. *Comput. Intell. Neurosci.* **2011**, *2011*, 9. [CrossRef]
64. Pascual-Marqui, R.D. Discrete, 3D distributed, linear imaging methods of electric neuronal activity. Part 1: Exact, zero error localization. *arXiv* **2007**, arXiv:0710.3341.
65. Pascual-Marqui, R.D.; Michel, C.M.; Lehmann, D. Low resolution electromagnetic tomography: A new method for localizing electrical activity in the brain. *Int. J. Psychophysiol.* **1994**, *18*, 49–65. [CrossRef]
66. Aoki, Y.; Ishii, R.; Pascual-Marqui, R.D.; Canuet, L.; Ikeda, S.; Hata, M.; Imajo, K.; Matsuzaki, H.; Musha, T.; Asada, T.; et al. Detection of EEG-resting state independent networks by eLORETA-ICA method. *Front. Human Neurosci.* **2015**, *9*, 31. [CrossRef] [PubMed]
67. Maris, E.; Oostenveld, R. Nonparametric statistical testing of EEG-and MEG-data. *J. Neurosci. Methods* **2007**, *164*, 177–190. [CrossRef] [PubMed]
68. Karniski, W.; Blair, R.C.; Snider, A.D. An exact statistical method for comparing topographic maps, with any number of subjects and electrodes. *Brain Topogr.* **1994**, *6*, 203–210. [CrossRef] [PubMed]
69. Bögels, S.; Barr, D.J.; Garrod, S.; Kessler, K. Conversational Interaction in the Scanner: Mentalizing during Language Processing as Revealed by MEG. *Cereb. Cortex* **2014**. [CrossRef]
70. Wang, H.; Callaghan, E.; Gooding-Williams, G.; McAllister, C.; Kessler, K. Rhythm makes the world go round: An MEG-TMS study on the role of right TPJ theta oscillations in embodied perspective taking. *Cortex* **2016**, *75*, 68–81. [CrossRef]
71. Fieldtrip. How Can I Test An Interaction Effect Using Cluster-Based Permutation Tests? Available online: http://www.fieldtriptoolbox.org/faq/how_can_i_test_an_interaction_effect_using_cluster-based_permutation_tests (accessed on 8 December 2019).
72. Gola, M.; Kaminski, J.; Brzezicka, A.; Wrobel, A. Beta band oscillations as a correlate of alertness—Changes in aging. *Int. J. Psychophysiol.* **2012**, *85*, 62–67. [CrossRef]
73. Gola, M.; Magnuski, M.; Szumska, I.; Wrobel, A. EEG beta band activity is related to attention and attentional deficits in the visual performance of elderly subjects. *Int. J. Psychophysiol.* **2013**, *89*, 334–341. [CrossRef]
74. Donoghue, J.P.; Sanes, J.N.; Hatsopoulos, N.G.; Gaál, G. Neural discharge and local field potential oscillations in primate motor cortex during voluntary movements. *J. Neurophysiol.* **1998**, *79*, 159–173. [CrossRef]
75. Farmer, S.F. Rhythmicity, synchronization and binding in human and primate motor systems. *J. Physiol.* **1998**, *509*, 3–14. [CrossRef]
76. Gross, J.; Schmitz, F.; Schnitzler, I.; Kessler, K.; Shapiro, K.; Hommel, B.; Schnitzler, A. Anticipatory control of long-range phase synchronization. *Eur. J. Neurosci.* **2006**, *24*, 2057–2060. [CrossRef] [PubMed]
77. Kühn, A.A.; Williams, D.; Kupsch, A.; Limousin, P.; Hariz, M.; Schneider, G.H.; Yarrow, K.; Brown, P. Event-related beta desynchronization in human subthalamic nucleus correlates with motor performance. *Brain* **2004**, *127*, 735–746. [CrossRef] [PubMed]
78. Griffin, I.C.; Miniussi, C.; Nobre, A.C. Orienting attention in time. *Front. Biosci.* **2001**, *6*, 660–671. [CrossRef]
79. Woodrow, H. The measurement of attention. *Psychol. Monogr.* **1914**, *17*, 1–158. [CrossRef]
80. Correa, A.; Nobre, A.C. Spatial and temporal acuity of visual perception can be enhanced selectively by attentional set. *Exp. Brain Res.* **2008**, *189*, 339–344. [CrossRef]
81. Posner, M.I.; Cohen, Y. Components of visual orienting. *Atten. Perform. X Control Lang. Process.* **1984**, *32*, 531–556.
82. Posner, M.I.; Rafal, R.D.; Choate, L.S.; Vaughan, J. Inhibition of return: Neural basis and function. *Cogn. Neuropsychol.* **1985**, *2*, 211–228. [CrossRef]
83. Neider, M.B.; Kramer, A.F. Older adults capitalize on contextual information to guide search. *Exp. Aging Res.* **2011**, *37*, 539–571. [CrossRef]
84. McLaughlin, P.M.; Murtha, S.J.E. The Effects of Age and Exogenous Support on Visual Search Performance. *Exp. Aging Res.* **2010**, *36*, 325–345. [CrossRef]
85. Watson, D.G.; Maylor, E.A. Aging and visual marking: Selective deficits for moving stimuli. *Psychol. Aging* **2002**, *17*, 321–339. [CrossRef]
86. Gamboz, N.; Zamarian, S.; Cavallero, C. Age-Related Differences in the Attention Network Test (ANT). *Exp. Aging Res.* **2010**, *36*, 287–305. [CrossRef] [PubMed]
87. Deiber, M.-P.; Ibanez, V.; Missonnier, P.; Rodriguez, C.; Giannakopoulos, P. Age-associated modulations of cerebral oscillatory patterns related to attention control. *Neuroimage* **2013**, *82*, 531–546. [CrossRef] [PubMed]

88. Zanto, T.P.; Pan, P.; Liu, H.; Bollinger, J.; Nobre, A.C.; Gazzaley, A. Age-Related Changes in Orienting Attention in Time. *J. Neurosci.* **2011**, *31*, 12461–12470. [CrossRef] [PubMed]
89. Gajewski, P.D.; Ferdinand, N.K.; Kray, J.; Falkenstein, M. Understanding sources of adult age differences in task switching: Evidence from behavioral and ERP studies. *Neurosci. Biobehav. Rev.* **2018**, *92*, 255–275. [CrossRef]
90. Callaghan, E.; Holland, C.; Kessler, K. Flexible allocation of attention in time or space across the life span: Theta and alpha oscillatory signatures of age-related decline and compensation as revealed by MEG. *Biorxiv* **2018**, 461020. [CrossRef]
91. Andrews, E.C.; Westerman, S.J. Age differences in simulated driving performance: Compensatory processes. *Accid. Anal. Prev.* **2012**, *45*, 660–668. [CrossRef]
92. Jefferies, L.; Roggeveen, A.; Enns, J.; Bennett, P.; Sekuler, A.; Di Lollo, V. On the time course of attentional focusing in older adults. *Psychol. Res.* **2015**, *79*, 28–41. [CrossRef]
93. Summala, H.; Lamble, D.; Laakso, M. Driving experience and perception of the lead car's braking when looking at in-car targets. *Accid. Anal. Prev.* **1998**, *30*, 401–407. [CrossRef]
94. Bao, S.; Boyle, L.N. Age-related differences in visual scanning at median-divided highway intersections in rural areas. *Accid. Anal. Prev.* **2009**, *41*, 146–152. [CrossRef]
95. Lopez, M.E.; Aurtenetxe, S.; Pereda, E.; Cuesta, P.; Castellanos, N.P.; Bruna, R.; Niso, G.; Maestu, F.; Bajo, R. Cognitive reserve is associated with the functional organization of the brain in healthy aging: A MEG study. *Front. Aging Neurosci.* **2014**, *6*. [CrossRef]
96. Anstey, K.; Christensen, H. Education, activity, health, blood pressure and apolipoprotein E as predictors of cognitive change in old age: A review. *Gerontology* **2000**, *46*, 163–177. [CrossRef] [PubMed]
97. Fratiglioni, L.; Paillard-Borg, S.; Winblad, B. An active and socially integrated lifestyle in late life might protect against dementia. *Lancet Neurol.* **2004**, *3*, 343–353. [CrossRef]
98. Roldán-Tapia, L.; García, J.; Cánovas, R.; León, I. Cognitive Reserve, Age, and Their Relation to Attentional and Executive Functions. *Appl. Neuropsychol.: Adult* **2012**, *19*, 2–8. [CrossRef] [PubMed]
99. Marottoli, R.A.; Leon, C.F.M.; Glass, T.A.; Williams, C.S.; Cooney, L.M.; Berkman, L.F.; Tinetti, M.E. Driving cessation and increased depressive symptoms: Prospective evidence from the New Haven EPESE. *J. Am. Geriatr. Soc.* **1997**, *45*, 202–206. [CrossRef] [PubMed]
100. Ragland, D.R.; Satariano, W.A.; MacLeod, K.E. Driving cessation and increased depressive symptoms. *J. Gerontol. Ser. A Biol. Sci. Med. Sci.* **2005**, *60*, 399–403. [CrossRef] [PubMed]
101. Windsor, T.D.; Anstey, K.J.; Butterworth, P.; Luszcz, M.A.; Andrews, G.R. The role of perceived control in explaining depressive symptoms associated with driving cessation in a longitudinal study. *Gerontologist* **2007**, *47*, 215–223. [CrossRef]
102. Paulo, A.C.; Sampaio, A.; Santos, N.C.; Costa, P.S.; Cunha, P.; Zihl, J.; Cerqueira, J.; Palha, J.A.; Sousa, N. Patterns of Cognitive Performance in Healthy Ageing in Northern Portugal: A Cross-Sectional Analysis. *PLoS ONE* **2011**, *6*, e24553. [CrossRef]
103. De Felice, S.; Holland, C.A. Intra-Individual Variability Across Fluid Cognition Can Reveal Qualitatively Different Cognitive Styles of the Aging Brain. *Front. Psychol.* **2018**, *9*. [CrossRef]

Article

Optimising Cognitive Enhancement: Systematic Assessment of the Effects of tDCS Duration in Older Adults

Claire J. Hanley *, Sophie L. Alderman and Elinor Clemence

Department of Psychology, Swansea University, Singleton Park, Swansea SA2 8PP, UK; sophiealderman24@gmail.com (S.L.A.); elinorclemence@gmail.com (E.C.)

* Correspondence: c.j.hanley@swansea.ac.uk

Received: 24 April 2020; Accepted: 14 May 2020; Published: 16 May 2020

Abstract: Transcranial direct current stimulation (tDCS) has been shown to support cognition and brain function in older adults. However, there is an absence of research specifically designed to determine optimal stimulation protocols, and much of what is known about subtle distinctions in tDCS parameters is based on young adult data. As the first systematic exploration targeting older adults, this study aimed to provide insight into the effects of variations in stimulation duration. Anodal stimulation of 10 and 20 min, as well as a sham-control variant, was administered to dorsolateral prefrontal cortex. Stimulation effects were assessed in relation to a novel attentional control task. Ten minutes of anodal stimulation significantly improved task-switching speed from baseline, contrary to the sham-control and 20 min variants. The findings represent a crucial step forwards for methods development, and the refinement of stimulation to enhance executive function in the ageing population.

Keywords: transcranial direct current stimulation; non-invasive brain stimulation; stimulation duration; aging; neural plasticity; attentional control

---

1. Introduction

Age-related neurochemical, structural, and functional brain changes are most pronounced in prefrontal regions and produce deficits in response inhibition [1,2], which drastically impact daily living, limiting personal safety, independence, and quality of life [3–5]. Such concerns represent a prominent societal challenge as life expectancy increases [6,7]. As pharmacological interventions have been largely ineffective [8–11], it is imperative that innovative strategies are developed to reduce the incidence of cognitive deficits.

In recent years, transcranial direct current stimulation (tDCS) has gained interest as a non-invasive and cost-effective method of enhancing cognition, due to its observed neuromodulatory effects on plasticity [12–14], particularly deficient neurotransmission [15], which is reported to underlie the presence of cognitive decline on neuropsychological tests [16]. Consequently, the existing evidence signals that the use of tDCS would be highly advantageous in minimising executive deficits. The vast majority of studies have focused on aspects of memory, where some success in enhancing the efficiency of working memory has been described in cognitively healthy older adults [17–19]. However, the comparatively limited literature on attentional control means mixed results are even more difficult to interpret [20–22]. This discrepancy may be accounted for by subtle variations in stimulation protocols, such that systematic evaluation of individual parameters is necessary to determine optimal results.

Studies in young adults have highlighted the non-linearity of variations in key stimulation parameters, such as current intensity [23–25]. It is not known whether the older population also demonstrate this pattern of results; however, the incidence of age-related brain atrophy likely necessitates the use of distinct protocols, compared to those that are effective in young adults [26–28]. Little is known about differences in duration, a crucial variable in relation to the induction of neuroplastic effects [29,30]. A computational modelling study [31] noted reductions in the peak electric fields generated in older adult participants, with the authors suggesting that longer durations of stimulation (than those typically used in conjunction with young adults) may prevent this. Therefore, stimulation of 20 min in length may be ideal where modulations of neuroplasticity are delayed due to diminished integrity of existing mechanisms [32] but, to date, this has not been formally tested.

The aim of this current study was to provide vital insight into the effects of duration for the purpose of further developing the use of tDCS, and refining stimulation protocols, specifically designed for older adults. This was achieved by assessing participants' task-switching speed, following anodal stimulation of 10 and 20 min, alongside that obtained during a sham-control condition. In line with the consensus in the literature, it was anticipated that task-switching speed would be enhanced after receiving active tDCS for the longer duration.

2. Materials and Methods

2.1. Subjects

In total, 40 participants, aged 60–75 years (67.05 ± 5.21, 20 females) were recruited to take part in the study. Prior to recruitment, all participants were asked to complete a screening form. Those with safety screening contraindications were excluded from the study. Contraindications included history of neurological (e.g., seizures, stroke) and/or psychiatric conditions (e.g., anxiety, depression), head trauma, concussion, and surgical implants (e.g., neurostimulator, pacemaker, cochlear implant). Individuals who had been prescribed medication designed to directly influence cortical excitation/inhibition (e.g., gabapentin for nerve pain), which may interfere with the emergence of tDCS effects, were also excluded [33]. All participants had corrected-to-normal vision, and scored in the normal range on the Montreal Cognitive Assessment (MoCA) [34] (27.80 ± 1.18). Participants gave written informed consent prior to taking part in the study. Procedures were carried out with the approval of the local ethics committee (Department of Psychology, Swansea University).

2.2. Task-Switching Paradigm

The task used was identical to that outlined in Hanley and Tales (2019) [22]. The Swansea Test of Attentional Control (STAC) is a complex task-switching paradigm, comprising selective attention, task monitoring, and response inhibition components (Figure 1). Use of a flexible algorithm designed to track performance (Parameter Estimation by Sequential Testing (PEST) [35]) calibrates speed on the basis of prior responses. PEST facilitates completion of the task within the bounds of an individual's capabilities and ensures that participants are able to respond successfully while not compromising on task difficulty, thereby, making the STAC ideal for use with older adult participants.

Participants were required to remain vigilant throughout the task in order to update the search criteria. The target changed every 12 s, resulting in approximately 25 targets per experimental run of 300 s. Speed (measured in symbols per minute per column; abbreviated to 'spm') was adjusted to maintain accuracy around a 75% correct criterion, using the PEST algorithm. Task speed began at 41 spm and increased or decreased in line with accuracy, such that task difficulty corresponded with performance. The participants' threshold is the speed at which the task is performed at the end of the test (referred to as final speed), whereby higher values represent superior performance.

Figure 1. The Swansea Test of Attentional Control (STAC) task. A target is identified within the 3 × 3 matrix of symbols (right). When a matching symbol appears amongst the three columns of the search array that scroll up the screen (left), participants press the spacebar as the symbol crosses behind the red line (as depicted in Hanley and Tales, 2019, [22]).

2.3. Transcranial Direct Current Stimulation

With the exception of duration, which was varied in the present study, parameters were identical to those outlined here [22]. Anodal stimulation of 10 and 20 min (1.5 mA), as well as a sham-control variant (10 min), was administered via 25 cm^2 electrodes positioned in a bihemispheric montage designed to target dorsolateral prefrontal cortex (dlPFC; F3/F4). In line with the available literature [26–28], the electrode size was smaller and stimulation intensity was greater than that typically used in conjunction with younger adults, in order to increase the focality of the current and compensate for increases in cerebrospinal fluid (CSF) observed in the ageing brain.

2.4. Experimental Procedure

Each participant received the three variants of stimulation (Sham, Active10, Active20) in a counterbalanced order, determined by a random sequence generator, with 7 days between subsequent sessions. Prior to acquiring the baseline data, participants executed the task for approximately 5 target changes to gain experience with the paradigm. Baseline data was acquired prior to stimulation (at the onset of their first session), which was compared to post-stimulation performance measures. Stimulation was administered while participants watched a nature documentary. After stimulation, they were asked to complete an adverse effects questionnaire (AEQ) to determine the presence and severity of stimulation side-effects.

2.5. Data Analysis

Data from all 40 participants was entered into statistical analysis using SPSS for Windows software (version 22; IBM, New York). Repeated-measures ANOVAs were used to assess differences relating to the AEQ data across sessions. To identify distinctions in task performance, a one-way, repeated-measures ANOVA was conducted on the STAC final speed data from each acquisition (Baseline, Sham, Active10, Active20). An alpha level of 0.05 was used to determine significance. Bonferroni corrected, post-hoc tests were conducted to investigate the main effect (significant differences from baseline in each of the three experimental conditions, with an adjusted alpha of 0.017).

3. Results

3.1. Adverse Effects Questionnaire

Participants reported mild–moderate side effects of stimulation. These reports were consistent across each of the three sessions (producing non-significant differences in tingling, burning, and concentration; $p > 0.05$).

3.2. Task-Switching Speed

A repeated measures ANOVA revealed a significant difference in task performance across conditions ($F(3,117) = 3.016$, $p = 0.033$, $\eta p^2 = 0.072$). Post-hoc t-tests established that this difference was driven by superior task speed in the Active10 compared to baseline condition ($t(39) = -4.227$, $p < 0.001$) (Figure 2). This result corresponded to a moderate effect size of 0.494 (Cohen's d; see [36]). Comparisons between baseline and sham ($t(39) = -1.059$, $p = 0.296$) and baseline and Active20 ($t(39) = -1.865$, $p = 0.070$) conditions were statistically non-significant.

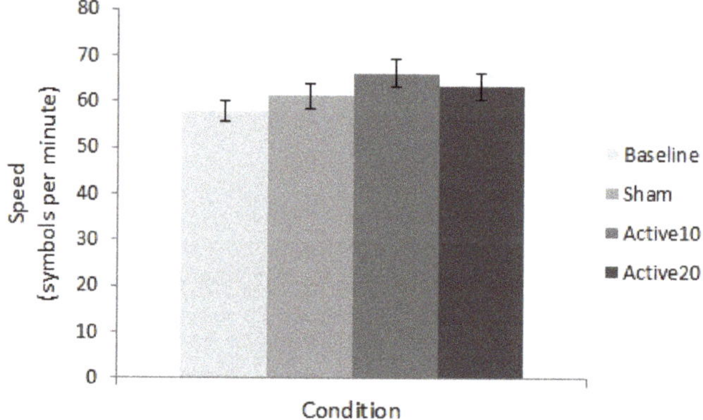

Figure 2. STAC final speed. Mean speed values for all conditions (baseline, sham, Active10, Active20) illustrate superior task performance following 10 min of anodal tDCS. Error bars represent ±1 standard error.

4. Discussion

The aim of the present study was to investigate the influence of variations in tDCS parameters as applied to older adults, specifically, by focusing on stimulation duration. When compared to the baseline condition, task-switching speed was significantly enhanced following 10 min of active stimulation; a result which assists in strengthening the limited evidence base in favour of using tDCS to enhance attentional control [21,22]. Neurochemical and/or functional imaging measures would be required to confirm the neurobiological underpinnings of the effect; however, in line with the dominant explanation for tDCS-induced enhancements, it is speculated that performance was facilitated by improved prefrontal network connectivity via the modification of NMDA/GABA receptor response, essential for promoting synaptic plasticity [29,30,37]. Where previous research has failed to establish desirable modulations of attentional control in older adults [20], this may be due to a lack of consideration of such neurobiological mechanisms. Accordingly, the aforementioned study by Boggio et al. directly replicated a tDCS protocol designed for young adults with an older adult sample. While the authors state that this decision stemmed from the aim of comparing performance, it nonetheless highlights a lack of appropriate study design where the populations in question inevitably differ in relation to key neural characteristics. In contrast, in the present study, the selected parameters enhance the biological plausibility of the rationale [38], by reflecting knowledge of age-related brain changes in the context of stimulation [26–28].

At the onset of the study, it was predicted that active stimulation, of 20 min in length, would be required to enhance task-switching ability. Conversely, 10 min of anodal tDCS significantly improved STAC final speed, thus challenging the suggestion that longer durations of stimulation are necessary to improve cognition in older adults [27,31,39–41]. To date, the limited available literature demonstrates the emergence of cognitive enhancement following 15+ min of active tDCS [13,42,43]. Such stimulation durations are said to compensate for excess CSF, characteristic of the ageing brain, which has been reported to reduce the focality of the current [26,28]. This includes our previous work that adopted a 20 min stimulation protocol, in which improved task-switching speed was established after three subsequent sessions [22]. Therefore, the observation of a single session improvement is equally intriguing given findings of delayed neuroplastic effects in older adults [32], which we had presumed would largely prevent this population from demonstrating an acute response to stimulation.

Intra-individual variability and non-linear responses to stimulation may account for the emergence of a significant effect at 10 min [44,45]. Accordingly, subtle changes in protocols can have marked effects on the resulting outcomes, hence the need for systematic evaluation of parameters. This implies that individuals have an optimal threshold, attributed to homeostatic constraints on neurobiological circuits to prevent over-excitation of calcium channels and NMDA receptors [46,47]. This effect is readily observed where stimulation is delivered at various intensities [48,49]. These studies demonstrate reliability between subsequent repeats of the same protocol yet assert that higher current strengths are not always necessary to produce modulations of excitability. Similarly, stimulation that is insufficient to fulfil an individual's optimal threshold may propagate deficient calcium transmission. For example, while increased intracellular calcium is integral for LTP, exceeding optimal levels will activate potassium channels and induce hyperpolarisation, forcing the cell population into a state of LTD or the so called 'no man's land' [50]. This is likely to result in the abolishment of expected neuromodulatory effects [46,51]. Furthermore, in the context of cathodal stimulation, typical inhibitory effects have been shown to be reversed, generating excitation at heightened intensities due to excessive stimulation and habituation of potassium channel response [23]. Cathodal stimulation is likely to diminish performance in older adults in contexts where the anodal polarity has been shown to be successful [52], and performance enhancement was a key objective of the present research. However, it would be interesting to determine whether an equivalent pattern of performance could be produced following inhibitory stimulation, which could potentially aid our interpretation of results.

Given the similarity in methodology, it is not likely that the task or elements of the stimulation protocol (beyond duration) contributed to the distinction between our studies, with regard to the generation of a significant effect following a single session of tDCS. Instead, subtle differences in the samples may account for the disparity in findings. Older adults are a particularly heterogeneous group and individual differences in tDCS response, like those found in conjunction with other non-invasive stimulation methods, are projected to account for approximately 40–50% of variance in outcomes [53,54]. Factors such as genetic variance (e.g., in relation to the regulation of plasticity; Brain-Derived Neurotrophic Factor (BDNF)) may be particularly relevant in the context of older adults, as those who are Val66Met carriers have been established to achieve maximal benefits following longer stimulation durations (20 compared to 10 min) [55]. The Val66Met polymorphism has been linked to a reduction in glutamatergic transmission [56,57], such that longer stimulation durations are required to induce neuroplastic effects. Therefore, variations in the capacity of an individual to modulate plasticity may have profound effects on stimulation outcomes.

Plasticity is known to decline with age [58] and, for this reason, it is likely desirable to keep age ranges fairly narrow when conducting tDCS research with older adults. This may be an additional reason why some stimulation studies fail to establish beneficial effects in the ageing population (for example [20], in which participants ranged from 50 to 85 years). Between our studies, there was a slight difference in age range, whereby our previous study recruited individuals aged 54–75, compared to 60–75 years in this instance, but both sets of participants had a similar average age (66.5 and 67 years, respectively). Therefore, age *per se* is unlikely to have been a defining factor. Furthermore, average MoCA scores between cohorts were also similarly high (28.2 and 27.8, respectively), signalling that variation in general cognitive function was also an unlikely cause. It is important to note, however, that MoCA score is not directly related to the incidence of frontal atrophy as may be expected [16], suggesting that identical test scores do not equate to similar patterns of atrophy. Distinctions in brain anatomy are likely independent of neuropsychological test outcome; such that where samples perform equally well on a standard cognitive measure, this does not mean they are identical from the perspective of neural change. Consequently, variation in results may be attributed to individual differences in brain structure and function that are commonly associated with older adults [59–61].

Anecdotally, many of the participants in the present study were still in employment and reported engaging in regular physical activity, lifestyle factors that mediate age-related decline in grey matter volume and white matter integrity [62,63]. The fact that repeated stimulation worked in the context of the previous study suggests the incidence of greater neural changes, explaining the need for lengthy stimulation, across multiple sessions, in order to alter plasticity and resulting cognitive performance [32]. Given the presence of a more 'youth-like' sample than that previously recruited, the neuroplastic mechanisms we sought to strengthen with tDCS may have still been largely intact in the present group, hence why they benefitted from a single session protocol. Without individual anatomical data, we are unable to confirm these differences in neuroanatomy; however, in young adults, long stimulation durations are not necessary to produce cognitive change [23,46]. This is also likely to be the case in the context of older adults, who recruit typical patterns of brain activity and still display hemispheric specialisation [59–61]. We intend to investigate this in the future by profiling participants in relation to several structural and functional neuroimaging metrics, as well as individual differences in lifestyle factors, because the integrity of the brain could be key in establishing the optimal duration of stimulation.

With the acquisition of neuroimaging data, computational modelling would also be possible, similar to that which has established changes in the effects of stimulation in the context of increased CSF [26,28]. A recent study has produced additional evidence to suggest that patterns of atrophy contribute to the amount of current reaching the cortex [64], which highlights the need for further systematic evaluations of approaches to compensate for such shortcomings (e.g., incrementally increasing the intensity of stimulation). Ultimately, generating a biologically plausible forward model to establish the likely outcome of stimulation, given the neuroanatomical status of an individual, could prove to be an incredibly valuable way of enhancing the validity of subsequent research [65]. Specifically, such a model could assist in the development of stimulation protocols to enhance cognition in older adults by providing crucial insight into optimal intensity and advantageous electrode positioning [66–68].

Incorporating online stimulation, during the task, may also enhance the effectiveness of tDCS. Meta-analyses highlight the benefits of online protocols in older adults due to age-related deficits in plasticity induction [69,70] (although this may largely apply to the motor domain, as opposed to cognition [41]). Nonetheless, it may be advantageous to isolate potential differences in the state-dependency of effects. While such meta-analyses converge on the consensus that tDCS is able to benefit cognitive performance, there is divergence between subtypes, such that it would be useful for studies to be able to compare across domains. This current study was designed to provide further insight into performance enhancement in the under-represented area of attentional control; however, incorporating a working memory task into the procedure would have allowed for a comparison of the findings to a wider range of past literature. In future, an N-back task [71] could act as a valid control measure, for example, because cognitive load can be modulated to parallel the complexity of the STAC. Such an addition would provide the basis for cross-domain inferences on the potential for global cognitive benefits, which could translate to improved function in aspects of daily life [72].

Lastly, with regard to methodological limitations of the stimulation protocol, it should be noted that the single sham session of 10 min prevented complete blinding. For this reason, the study is regarded as a 'partial blind' because, while the 20 min stimulation would have been discernible, the nature of the two 10 min sessions was unknown (to both participants and the researchers), as codes were used to execute stimulation. Although differences in duration are likely more obvious, participants can detect subtleties in current strength (particularly where higher intensities are used [73]), yet researchers commonly use a single sham session in the context of systematic investigations of intensity [19,23,24]. This is most likely due to the already high number of sessions required to conduct systematic evaluation studies (both an inherent strength and weakness of a within-subject experimental design), which focus on the influence of variations in active stimulation. One particular study has used this rationale to omit a sham control condition altogether [44]. While this is likely not advisable, the consensus remains that no specific approach to sham stimulation appears to be any more rigorous than another (including repeated sham conditions) [74]. Evidently, there is still much to be learned about the intricacies of control stimulation, particularly in the context of the older adult population.

In conclusion, advances in our understanding of tDCS effects in the context of older adulthood are very much dependent on methodological development and continued research. These results attest to the safety and tolerability of tDCS in older adults [75] and provide a framework within which to continue testing existing mechanistic assumptions, relating to key parameters, and build momentum in advancing towards flexible and feasible strategies to target age-related changes in cognition. Where this can be achieved, progress towards maintaining executive function in the ageing population is likely to translate to respective benefits in tasks of daily function, an increasingly important consideration as life expectancy continues to rise.

Author Contributions: Conceptualisation, C.J.H.; methodology, C.J.H., S.L.A., and E.C.; validation, C.J.H.; formal analysis, C.J.H., S.L.A., and E.C.; investigation, S.L.A. and E.C.; resources, C.J.H.; data curation, C.J.H., S.L.A., and E.C.; writing—original draft preparation, C.J.H.; writing—review and editing, C.J.H., S.L.A., and E.C.; visualisation, C.J.H.; supervision, C.J.H.; project administration, C.J.H.; funding acquisition, C.J.H. All authors have read and agreed to the published version of the manuscript.

Funding: This research received no external funding. The research was funded by the Department of Psychology, Swansea University. This research did not receive any specific grant from funding agencies in the public, commercial, or not-for-profit sectors.

Acknowledgments: The authors would like to thank the Department of Psychology, Swansea University, for funding the research, as well as the Dementia Research Group (College of Human and Health Sciences, Swansea University) and the associated database of volunteers who reviewed, and took part in the study, respectively.

Conflicts of Interest: The authors declare no conflict of interest.

References

1. Sylvester, C.-Y.C.; Wager, T.; Lacey, S.C.; Hernandez, L.; Nichols, T.E.; E Smith, E.; Jonides, J. Switching attention and resolving interference: fMRI measures of executive functions. *Neuropsychologia* **2003**, *41*, 357–370. [CrossRef]
2. Cieslik, E.C.; Mueller, V.I.; Eickhoff, C.R.; Langner, R.; Eickhoff, S.B. Three key regions for supervisory attentional control: Evidence from neuroimaging meta-analyses. *Neurosci. Biobehav. Rev.* **2015**, *48*, 22–34. [CrossRef] [PubMed]
3. Logsdon, R.G.; Gibbons, L.E.; McCurry, S.M.; Teri, L. Assessing quality of life in older adults with cognitive impairment. *Psychosom. Med.* **2002**, *64*, 510–519. [CrossRef] [PubMed]
4. Emilien, G.; Durlach, C.; Minaker, K.L.; Winblad, B.; Gauthier, S.; Maloteaux, J.-M. Mild Cognitive Impairment. *Lancet* **2006**, *367*, 1262–1270.
5. Shimada, H.; Makizako, H.; Doi, T.; Lee, S.; Tsutsumimoto, K.; Hotta, R.; Bae, S.; Nakakubo, S.; Harada, K.; Suzuki, T. Impact of cognitive frailty on daily activities in older persons. *J. Nutr. Health Aging* **2016**, *20*, 729–735. [CrossRef]
6. Christensen, K.; Doblhammer, G.; Rau, R.; Vaupel, J.W. Ageing populations: The challenges ahead. *Lancet* **2009**, *374*, 1196–1208. [CrossRef]

7. Howdon, D.; Rice, N. Health care expenditures, age, proximity to death and morbidity: Implications for an ageing population. *J. Health Econ.* **2018**, *57*, 60–74. [CrossRef]
8. Whalley, L.J.; Deary, I.J.; Appleton, C.L.; Starr, J.M. Cognitive reserve and the neurobiology of cognitive aging. *Ageing Res. Rev.* **2004**, *3*, 369–382. [CrossRef]
9. Fritsch, T.; McClendon, M.J.; Smyth, K.A.; Lerner, A.J.; Friedland, R.P.; Larsen, J.D. Cognitive functioning in healthy aging: The role of reserve and lifestyle factors early in life. *Gerontologist* **2007**, *47*, 307–322. [CrossRef]
10. Kane, M.R.L.; Butler, P.M.; Fink, M.H.A. Interventions to prevent age-related cognitive decline, mild cognitive impairment, and clinical Alzheimer's-type dementia. *J. Comp. Eff. Rev.* **2017**, *188*. [CrossRef]
11. Northey, J.M.; Cherbuin, N.; Pumpa, K.L.; Smee, D.J.; Rattray, B. Exercise interventions for cognitive function in adults older than 50: A systematic review with meta-analysis. *Br. J. Sports Med.* **2018**, *52*, 154–160. [CrossRef] [PubMed]
12. Pisoni, A.; Mattavelli, G.; Papagno, C.; Rosanova, M.; Casali, A.G.; Romero Lauro, L.J. Cognitive enhancement induced by anodal tDCS drives circuit-specific cortical plasticity. *Cereb. Cortex* **2018**, *28*, 1132–1140. [CrossRef] [PubMed]
13. Meinzer, M.; Lindenberg, R.; Antonenko, D.; Flaisch, T.; Flöel, A. Anodal transcranial direct current stimulation temporarily reverses age-associated cognitive decline and functional brain activity changes. *J. Neurosci.* **2013**, *33*, 12470–12478. [CrossRef] [PubMed]
14. Antonenko, D.; Nierhaus, T.; Meinzer, M.; Prehn, K.; Thielscher, A.; Ittermann, B.; Flöel, A. Age-dependent effects of brain stimulation on network centrality. *NeuroImage* **2018**, *176*, 71–82. [CrossRef]
15. Antonenko, D.; Schubert, F.; Bohm, F.; Ittermann, B.; Aydin, S.; Hayek, D.; Grittner, U.; Flöel, A. tDCS-induced modulation of GABA levels and resting-state functional connectivity in older adults. *J. Neurosci.* **2017**, *37*, 4065–4073. [CrossRef]
16. Porges, E.C.; Woods, A.J.; Edden, R.A.; Puts, N.A.; Harris, A.D.; Chen, H.; Garcia, A.M.; Seider, T.R.; Lamb, D.G.; Williamson, J.B.; et al. Frontal gamma-aminobutyric acid concentrations are associated with cognitive performance in older adults. *Biol. Psychiatry Cogn. Neurosci. Neuroimaging* **2017**, *2*, 38–44. [CrossRef]
17. Park, S.H.; Seo, J.H.; Kim, Y.H.; Ko, M.H. Long-term effects of transcranial direct current stimulation combined with computer-assisted cognitive training in healthy older adults. *Neuroreport* **2014**, *25*, 122–126. [CrossRef]
18. Jones, K.T.; Gözenman, F.; Berryhill, M.E. The strategy and motivational influences on the beneficial effect of neurostimulation: A tDCS and fNIRS study. *NeuroImage* **2015**, *105*, 238–247. [CrossRef]
19. Stephens, J.A.; Berryhill, M.E. Older adults improve on everyday tasks after working memory training and neurostimulation. *Brain Stimul.* **2016**, *9*, 553–559. [CrossRef]
20. Boggio, P.S.; Campanhã, C.; Valasek, C.A.; Fecteau, S.; Pascual-Leone, A.; Fregni, F. Modulation of decision-making in a gambling task in older adults with transcranial direct current stimulation. *Eur. J. Neurosci.* **2010**, *31*, 593–597. [CrossRef]
21. Harty, S.; Robertson, I.H.; Miniussi, C.; Sheehy, O.C.; Devine, C.A.; McCreery, S.; O'Connell, R.G. Transcranial direct current stimulation over right dorsolateral prefrontal cortex enhances error awareness in older age. *J. Neurosci.* **2014**, *34*, 3646–3652. [CrossRef] [PubMed]
22. Hanley, C.J.; Tales, A. Anodal tDCS improves attentional control in older adults. *Exp. Gerontol.* **2019**, *115*, 88–95. [CrossRef] [PubMed]
23. Batsikadze, G.; Moliadze, V.; Paulus, W.; Kuo, M.F.; Nitsche, M.A. Partially non-linear stimulation intensity-dependent effects of direct current stimulation on motor cortex excitability in humans. *J. Physiol.* **2013**, *591*, 1987–2000. [CrossRef] [PubMed]
24. Jamil, A.; Batsikadze, G.; Kuo, H.-I.; Labruna, L.; Hasan, A.; Paulus, W.; Nitsche, M.A. Systematic evaluation of the impact of stimulation intensity on neuroplastic after-effects induced by transcranial direct current stimulation. *J. Physiol.* **2017**, *595*, 1273–1288. [CrossRef] [PubMed]
25. Esmaeilpour, Z.; Marangolo, P.; Hampstead, B.M.; Bestmann, S.; Galletta, E.; Knotkova, H.; Bikson, M. Incomplete evidence that increasing current intensity of tDCS boosts outcomes. *Brain Stimul.* **2018**, *11*, 310–321. [CrossRef] [PubMed]
26. Laakso, I.; Tanaka, S.; Koyama, S.; De Santis, V.; Hirata, A. Inter-subject variability in electric fields of motor cortical tDCS. *Brain Stimul.* **2015**, *8*, 906–913. [CrossRef]
27. Tatti, E.; Rossi, S.; Innocenti, I.; Rossi, A.; Santarnecchi, E. Non-invasive brain stimulation of the aging brain: State of the art and future perspectives. *Ageing Res. Rev.* **2016**, *29*, 66–89. [CrossRef]

28. Mahdavi, S.; Towhidkhah, F. Alzheimer's Disease Neuroimaging Initiative. Computational human head models of tDCS: Influence of brain atrophy on current density distribution. *Brain Stimul.* **2018**, *11*, 104–107. [CrossRef]
29. Nitsche, M.A.; Fricke, K.; Henschke, U.; Schlitterlau, A.; Liebetanz, D.; Lang, N.; Henning, S.; Tergau, F.; Paulus, W. Pharmacological modulation of cortical excitability shifts induced by transcranial direct current stimulation in humans. *J. Physiol.* **2003**, *553*, 293–301. [CrossRef]
30. Stagg, C.J.; Nitsche, M.A. Physiological basis of transcranial direct current stimulation. *Neuroscientist* **2011**, *17*, 37–53. [CrossRef]
31. Thomas, C.; Datta, A.; Woods, A. Effect of aging on current flow due to transcranial direct current stimulation. *Brain Stimul.* **2017**, *10*, 469. [CrossRef]
32. Fujiyama, H.; Hyde, J.; Hinder, M.; Kim, S.-J.; McCormack, G.H.; Vickers, J.C.; Summers, J.J. Delayed plastic responses to anodal tDCS in older adults. *Front. Aging Neurosci.* **2014**, *6*, 115. [CrossRef] [PubMed]
33. McLaren, M.E.; Nissim, N.R.; Woods, A.J. The effects of medication use in transcranial direct current stimulation: A brief review. *Brain Stimul.* **2018**, *11*, 52–58. [CrossRef] [PubMed]
34. Nasreddine, Z.S.; Phillips, N.A.; Bédirian, V.; Charbonneau, S.; Whitehead, V.; Collin, I.; Cummings, J.L.; Chertkow, H.; Bédirian, V. The Montreal Cognitive Assessment, MoCA: A brief screening tool for mild cognitive impairment. *J. Am. Geriat. Soc.* **2005**, *53*, 695–699. [CrossRef] [PubMed]
35. Taylor, M.; Creelman, C.D. PEST: Efficient estimates on probability functions. *J. Acoust. Soc. Am.* **1967**, *41*, 782–787. [CrossRef]
36. Lakens, D. Calculating and reporting effect sizes to facilitate cumulative science: A practical primer for t-tests and ANOVAs. *Front. Psychol.* **2013**, *4*, 863. [CrossRef]
37. Filmer, H.L.; Dux, P.E.; Mattingley, J.B. Applications of transcranial direct current stimulation for understanding brain function. *Trends Neurosci.* **2014**, *37*, 742–753. [CrossRef]
38. Bestmann, S.; de Berker, A.O.; Bonaiuto, J. Understanding the behavioural consequences of noninvasive brain stimulation. *Trends Cogn. Sci.* **2015**, *19*, 13–20. [CrossRef]
39. Shekhawat, G.S.; Stinear, C.M.; Searchfield, G.D. Transcranial direct current stimulation intensity and duration effects on tinnitus suppression. *Neurorehabil. Neural Repair* **2013**, *27*, 164–172. [CrossRef]
40. Shekhawat, G.S.; Sundram, F.; Bikson, M.; Truong, D.Q.; De Ridder, D.; Stinear, C.; Welch, D.; Searchfield, G. Intensity, duration, and location of high-definition transcranial direct current stimulation for tinnitus relief. *Neurorehabil. Neural Repair* **2016**, *30*, 349–359. [CrossRef]
41. Summers, J.J.; Kang, N.; Cauraugh, J.H. Does transcranial direct current stimulation enhance cognitive and motor functions in the ageing brain? A systematic review and meta-analysis. *Ageing Res. Rev.* **2016**, *25*, 42–54. [CrossRef] [PubMed]
42. Antal, A.; Kincses, T.Z.; Nitsche, M.A.; Bartfai, O.; Paulus, W. Excitability changes induced in the human primary visual cortex by transcranial direct current stimulation: Direct electrophysiological evidence. *Invest. Ophthal. Vis. Sci.* **2004**, *45*, 702–707. [CrossRef] [PubMed]
43. Antal, A.; Paulus, W. Transcranial direct current stimulation and visual perception. *Perception* **2008**, *37*, 367–374. [CrossRef] [PubMed]
44. Tremblay, S.; Larochelle-Brunet, F.; Lafleur, L.P.; El Mouderrib, S.; Lepage, J.F.; Théoret, H. Systematic assessment of duration and intensity of anodal transcranial direct current stimulation on primary motor cortex excitability. *Eur. J. Neurosci.* **2016**, *44*, 2184–2190. [CrossRef]
45. Horvath, J.C.; Carter, O.; Forte, J.D. Transcranial direct current stimulation: Five important issues we aren't discussing (but probably should be). *Front. Syst. Neurosci.* **2014**, *8*, 2. [CrossRef]
46. Monte-Silva, K.; Kuo, M.-F.; Hessenthaler, S.; Fresnoza, S.; Liebetanz, D.; Paulus, W.; Nitsche, M.A. Induction of late LTP-like plasticity in the human motor cortex by repeated non-invasive brain stimulation. *Brain Stimul.* **2013**, *6*, 424–432. [CrossRef]
47. Krause, B.; Márquez-Ruiz, J.; Cohen Kadosh, R. The effect of transcranial direct current stimulation: A role for cortical excitation/inhibition balance? *Front. Human Neurosci.* **2013**, *7*, 602. [CrossRef]
48. Kidgell, D.; Daly, R.M.; Young, K.; Lum, J.; Tooley, G.; Jaberzadeh, S.; Zoghi, M.; Pearce, A.J. Different current intensities of anodal transcranial direct current stimulation do not differentially modulate motor cortex plasticity. *Neural Plast.* **2013**, *2013*, 603502. [CrossRef]
49. Bastani, A.; Jaberzadeh, S. Differential Modulation of Corticospinal Excitability by Different Current Densities of Anodal Transcranial Direct Current Stimulation. *PLoS ONE* **2013**, *8*, e72254. [CrossRef]

50. Lisman, J.E. Three Ca^{2+} levels affect plasticity differently: The LTP zone, the LTD zone and no man's land. *J. Physiol.* **2001**, *532*, 285. [CrossRef]
51. Opitz, A.; Paulus, W.; Will, S.; Antunes, A.; Thielscher, A. Determinants of the electric field during transcranial direct current stimulation. *NeuroImage* **2015**, *109*, 140–150. [CrossRef] [PubMed]
52. Bortoletto, M.; Pellicciari, M.C.; Rodella, C.; Miniussi, C. The interaction with task-induced activity is more important than polarization: A tDCS study. *Brain Stimul.* **2015**, *8*, 269–276. [CrossRef] [PubMed]
53. López-Alonso, V.; Fernández-del-Olmo, M.; Costantini, A.; Gonzalez-Henriquez, J.J.; Cheeran, B. Intra-individual variability in the response to anodal transcranial direct current stimulation. *Clin. Neurophysiol.* **2015**, *126*, 2342–2347. [CrossRef] [PubMed]
54. Wiethoff, S.; Hamada, M.; Rothwell, J.C. Variability in response to transcranial direct current stimulation of the motor cortex. *Brain Stimul.* **2014**, *7*, 468–475. [CrossRef] [PubMed]
55. Puri, R.; Hinder, M.R.; Fujiyama, H.; Gomez, R.; Carson, R.G.; Summers, J.J. Duration-dependent effects of the BDNF Val66Met polymorphism on anodal tDCS induced motor cortex plasticity in older adults: A group and individual perspective. *Front. Aging Neurosci.* **2015**, *7*, 107. [CrossRef] [PubMed]
56. Fritsch, B.; Reis, J.; Martinowich, K.; Schambra, H.M.; Ji, Y.; Cohen, L.G.; Lu, B. Direct current stimulation promotes BDNF-dependent synaptic plasticity: Potential implications for motor learning. *Neuron* **2010**, *66*, 198–204. [CrossRef] [PubMed]
57. Nathan, P.J.; Cobb, S.R.; Lu, B.; Bullmore, E.T.; Davies, C.H. Studying synaptic plasticity in the human brain and opportunities for drug discovery. *Curr. Opin. Pharmacol.* **2011**, *11*, 540–548. [CrossRef]
58. Fathi, D.; Ueki, Y.; Mima, T.; Koganemaru, S.; Nagamine, T.; Tawfik, A.; Fukuyama, H. Effects of aging on the human motor cortical plasticity studied by paired associative stimulation. *Clin. Neurophysiol.* **2010**, *121*, 90–93. [CrossRef]
59. Cabeza, R. Hemispheric asymmetry reduction in older adults: The HAROLD model. *Psychol. Aging* **2002**, *17*, 85. [CrossRef]
60. Reuter-Lorenz, P.A.; Park, D.C. How does it STAC up? Revisiting the scaffolding theory of aging and cognition. *Neuropsychol. Rev.* **2014**, *24*, 355–370. [CrossRef]
61. Cabeza, R.; Albert, M.; Belleville, S.; Craik, F.I.M.; Duarte, A.; Grady, C.L.; Lindenberger, U.; Nyberg, L.; Park, D.C.; Reuter-Lorenz, P.A.; et al. Maintenance, reserve and compensation: The cognitive neuroscience of healthy ageing. *Nat. Rev. Neurosci.* **2018**, *19*, 701–710. [CrossRef] [PubMed]
62. Seider, T.R.; Fieo, R.A.; O'Shea, A.; Porges, E.C.; Woods, A.J.; Cohen, R.A. Cognitively engaging activity is associated with greater cortical and subcortical volumes. *Front. Aging Neurosci.* **2016**, *8*, 94. [CrossRef] [PubMed]
63. Matura, S.; Fleckenstein, J.; Deichmann, R.; Engeroff, T.; Füzéki, E.; Hattingen, E.; Hellweg, R.; Lienerth, B.; Pilatus, U.; Schwarz, S.; et al. Effects of aerobic exercise on brain metabolism and grey matter volume in older adults: Results of the randomised controlled SMART trial. *Transl. Psychiatry* **2017**, *7*, e1172. [CrossRef] [PubMed]
64. Indahlastari, A.; Albizu, A.; O'Shea, A.; Forbes, M.A.; Nissim, N.R.; Kraft, J.N.; Evangelista, N.D.; Hausman, H.K.; Woods, A.J.; Initiative, A.D.N. Modeling transcranial electrical stimulation in the aging brain. *Brain Stimul.* **2020**, *13*, 664–674. [CrossRef] [PubMed]
65. Bikson, M.; Rahman, A.; Datta, A. Computational models of transcranial direct current stimulation. *Clin. EEG Neurosci.* **2012**, *43*, 176–183. [CrossRef]
66. Rampersad, S.M.; Janssen, A.; Lucka, F.; Aydin, U.; Lanfer, B.; Lew, S.; Wolters, C.; Stegeman, D.F.; Oostendorp, T.F. Simulating transcranial direct current stimulation with a detailed anisotropic human head model. *IEEE Trans. Neural Syst. Rehabil.* **2014**, *22*, 441–452. [CrossRef]
67. Mikkonen, M.; Laakso, I.; Tanaka, S.; Hirata, A. Cost of focality in TDCS: Interindividual variability in electric fields. *Brain Stimul.* **2020**, *13*, 117–124. [CrossRef]
68. Laakso, I.; Mikkonen, M.; Koyama, S.; Hirata, A.; Tanaka, S. Can electric fields explain inter-individual variability in transcranial direct current stimulation of the motor cortex? *Sci. Rep.* **2019**, *9*, 626. [CrossRef]
69. Morrison, J.H.; Baxter, M.G. The ageing cortical synapse: Hallmarks and implications for cognitive decline. *Nat. Rev. Neurosci.* **2012**, *13*, 240–250. [CrossRef]
70. Hsu, W.Y.; Ku, Y.; Zanto, T.P.; Gazzaley, A. Effects of noninvasive brain stimulation on cognitive function in healthy aging and Alzheimer's disease: A systematic review and meta-analysis. *Neurobiol. Aging* **2015**, *36*, 2348–2359. [CrossRef]

71. Brunoni, A.R.; Vanderhasselt, M.A. Working memory improvement with non-invasive brain stimulation of the dorsolateral prefrontal cortex: A systematic review and meta-analysis. *Brain Cogn.* **2014**, *86*, 1–9. [CrossRef] [PubMed]
72. Jones, K.T.; Stephens, J.A.; Alam, M.; Bikson, M.; Berryhill, M.E. Longitudinal neurostimulation in older adults improves working memory. *PLoS ONE* **2015**, *10*, e0121904. [CrossRef] [PubMed]
73. E O'Connell, N.; Cossar, J.; Marston, L.; Wand, B.M.; Bunce, D.; Moseley, G.L.; De Souza, L.H. Rethinking Clinical Trials of Transcranial Direct Current Stimulation: Participant and Assessor Blinding Is Inadequate at Intensities of 2mA. *PLoS ONE* **2012**, *7*, e47514. [CrossRef] [PubMed]
74. Fonteneau, C.; Mondino, M.; Arns, M.; Baeken, C.; Bikson, M.; Brunoni, A.R.; Poulet, E. Sham tDCS: A hidden source of variability? Reflections for further blinded, controlled trials. *Brain Stimul.* **2019**, *12*, 668–673. [CrossRef] [PubMed]
75. Goodwill, A.M.; Reynolds, J.; Daly, R.M.; Kidgell, D.J. Formation of cortical plasticity in older adults following tDCS and motor training. *Front. Aging Neurosci.* **2013**, *5*, 87. [CrossRef] [PubMed]